The publishing house tredition has created the series **TREDITION CLASSICS**. It contains classical literature works from over two thousand years. Most of these titles have been out of print and off the bookstore shelves for decades.

The book series is intended to preserve the cultural legacy and to promote the timeless works of classical literature. As a reader of a **TREDITION CLASSICS** book, the reader supports the mission to save many of the amazing works of world literature from oblivion.

The symbol of **TREDITION CLASSICS** is Johannes Gutenberg (1400 – 1468), the inventor of movable type printing.

With the series, tredition intends to make thousands of international literature classics available in printed format again – worldwide.

All books are available at book retailers worldwide in paperback and in hardcover. For more information please visit: www.tredition.com

tredition was established in 2006 by Sandra Latusseck and Soenke Schulz. Based in Hamburg, Germany, tredition offers publishing solutions to authors and publishing houses, combined with worldwide distribution of printed and digital book content. tredition is uniquely positioned to enable authors and publishing houses to create books on their own terms and without conventional manufacturing risks.

For more information please visit: www.tredition.com

Wonderful Balloon Ascents

F. (Fulgence) Marion

Imprint

This book is part of the TREDITION CLASSICS series.

Author: F. (Fulgence) Marion
Cover design: toepferschumann, Berlin (Germany)

Publisher: tredition GmbH, Hamburg (Germany)
ISBN: 978-3-8491-8876-4

www.tredition.com
www.tredition.de

Copyright:
The content of this book is sourced from the public domain.

The intention of the TREDITION CLASSICS series is to make world literature in the public domain available in printed format. Literary enthusiasts and organizations worldwide have scanned and digitally edited the original texts. tredition has subsequently formatted and redesigned the content into a modern reading layout. Therefore, we cannot guarantee the exact reproduction of the original format of a particular historic edition. Please also note that no modifications have been made to the spelling, therefore it may differ from the orthography used today.

BALLOON OF THE MARQUIS D'ARLANDES.

CONTENTS

PREFACE

List of Illustrations

BALLOONS AND AIR JOURNEYS.

PART I. THE CONQUEST OF THE SKIES.—1783.
Chapter I. Introduction.

Chapter II. Attempts in Ancient Times to Fly in the Air.

Chapter III. The Theory of Balloons.

Chapter IV. First Public Trial of the Balloon.

Chapter V. Second Experiment.

Chapter VI. Third Experiment.

Chapter VII. Fourth Experiment.

Chapter VIII. Men and Balloons.

Chapter IX. The First Aerial Voyage—Roziers and Arlandes.

Chapter X. The Second Arial Voyage.

PART II. The History of Aerostation from the Year 1783.
Chapter I.

Chapter II. Experiments and Studies—Blanchard at Paris—Guyton de Morveau at Dijon.

Chapter III.

Chapter IV.

Chapter V. First Aerial Voyage in England—Blanchard Crosses the Sea in a Balloon.

Chapter VI. Zambeccari's Perilous Trip Across the Adriatic Sea.

Chapter VII. Garnerin — Parachutes — Aerostation at Public Fetes.

Chapter VIII. Green's Great Journey Across Europe.

Chapter IX. The "Geant" Balloon.

Chapter X. The Necrology of Aeronautic

PART III. Scientific Experiments — Applications of Ballooning.

Chapter I. Experiments of Robertson, Lhoest, Saccarof, &c.

Chapter II. Ascent of M. Gay-Lussac Alone — Excursions of MM. Barral and Bixio.

Chapter III. Ascents of the Mssrs. Welsh, Glaisher and Coxwell.

Chapter IV. Balloons Made Useful in Warfare.

Advertisements in the back of the book

ASCENT OF THE 19TH SEPTEMBER, 1783, AT VERSAILLES.

Many other illustrations may be viewed in the List of Illustrations below

PREFACE

"Let posterity know, and knowing be astonished, that on the fifteenth day of September, 1784, Vincent Lunardi of Lucca, in Tuscany, the first aerial traveller in Britain, mounting from the Artillery Ground in London, and traversing the regions of the air for two hours and fifteen minutes, on this spot revisited the earth. In this rude monument for ages be recorded this wondrous enterprise successfully achieved by the powers of chemistry and the fortitude of man, this improvement in science which the great Author of all Knowledge, patronising by his Providence the inventions of mankind, hath graciously permitted, to their benefit and his own eternal glory."

The stone upon which the above inscription was carved, stands, or stood recently, near Collier's End, in the parish of Standon, Hertfordshire; and it will possibly afford the English reader a more accurate idea of the feelings with which the world hailed the discovery of the balloon than any incident or illustration drawn from the annals of a foreign country.

The work which we now introduce to our readers does not exaggerate the case when it declares that no discovery of modern times has aroused so large an amount of enthusiasm, has excited so many hopes, has appeared to the human race to open up so many vistas of enterprise and research, as that for which we are mainly indebted to the Brothers Montgolfier. The discovery or the invention of the balloon, however, was one of those efforts of genius and enterprise which have no infancy. It had reached its full growth when it burst upon the world, and the ninety years which have since elapsed have witnessed no development of the original idea. The balloon of today—the balloon in which Coxwell and Glaisher have made their perilous trips into the remote regions of the air—is in almost every respect the same as the balloon with which "the physician Charles," following in the footsteps of the Montgolfiers, astonished Paris in 1783. There are few more tantalising stories in the annals of invention than this. So much had been accomplished when Roziers made his first aerial voyage above the astonished capital of France that all the rest seemed easy. The new highway appeared to have been thrown open to the world, and the dullest imagination saw the air thronged with colossal chariots, bearing travellers in perfect safety, and with more than the speed of the eagle, from city to city, from country to country, reckless of all the obstacles—the seas, and rivers, and mountains—which Nature might have placed in the path of the wayfarer. But from that moment to the present the prospect which was thus opened up has remained a vision and nothing more. There are—as those who visited the Crystal Palace two years ago have reason to know—not a few men who still believe in the practicability of journeying by air. But, with hardly an exception, those few have abandoned all idea of utilising the balloon for this purpose. The graceful "machine" which astonished the world at its birth remains to this day as beautiful, and as useless for the purposes of travel, as in the first hour of its history. The day may come when some one more fortunate than the Montgolfiers may earn the Duke of Sutherland's offered reward by a successful flight from the Mall to the top of Stafford House; but when this comes to pass the balloon will have no share in the honour of the achievement. Not the less, however, is the story of this wonderful invention worthy of being recorded. It deserves a place in the history of human enterprise—if for nothing else—because of the daring courage which it has in so many cases brought to light. From the days of Roziers down to those of Coxwell, our aeronauts have fearlessly tempted dangers not less terrible than those which face the soldier as he enters the imminent deadly breach; and, as one of the chapters in this volume mournfully proves, not a few of their number have paid the penalty of their rash courage with their lives. All the more is it to be regretted that so little practical good has resulted from their labours and their sacrifices; and that so many of those who have perished in balloon voyages have done so whilst serving to better end than the amusement of a holiday crowd. There is, however, another aspect which makes at least the earlier history of the balloon well worth preserving. This is the influence which the invention had upon the generation which witnessed it. As these pages show, the people of Europe seem to have been absolutely intoxicated by the success of the Montgolfiers' discovery. There is something bitterly suggestive in our knowledge of this fact. Whilst pensions and honours and popular applause were being showered upon the inventors of the balloon, Watt was labouring unnoticed at his improvements of the steam-engine—a very prosaic affair compared with the gilded globe which Montgolfier had caused to rise from earth amidst the acclamations of a hundred thousand spectators, but one which had before it a somewhat different history to that of the more startling invention. England, when it remembers the story of the steam-engine, has little need to grudge France the honour of discovering the balloon. After all, however, Great Britain had its share in that discovery. The early observations of Francis Bacon and Bishop Wilkins paved the way for the later achievement, whilst it was our own Cavendish who discovered that hydrogen gas was lighter than air; and Dr. Black of Edinburgh, who first employed that gas to raise a globe in which it was contained from the earth. The Scotch professor, we are told, thought that the discovery which he made when he sent his little tissue-paper balloon from his lecture-table to the ceiling of his classroom, was of no use except as affording the means of making an interesting experiment. Possibly our readers, after they have perused this volume, may think that Dr Black was not after all so far wrong as people once imagined. Be this as it may, however, in these pages is the history of the balloon, and of the most memorable balloon voyages, and we comprehend the story to our readers not the less cordially that it comes from the land where the balloon had its birth.

London, January, 1870.

List of Illustrations

Click on any of the Illustrations

- 01. Lana's Flying Machine
- 02. Laurent de Guzman's Balloon
- 03. The Flying Man
- 04. Inflating Balloon with Hydrogen
- 05. The Parachute
- 06. Garnerin's Descent in a Parachute
- 07. The Brothers Montgolfier
- 08. Charles's Balloon on its way to the Champ de Mars
- 09. The Ascent of Charles's Balloon from the Champ de Mars
- 10. The Destruction of Charles's Balloon
- 11. Ascent of the 19th September, 1783, at Versailles
- 12. Balloon of the Marquis D'Arlandes
- 13. The Balloon of D'Arlandes crossing Paris
- 14. The First Aerial Voyage
- 15. Monsieur Charles and the Duke of Chartres
- 16. Bagnolet's Balloon 17. Le Flesselles
- 18. Blanchard's Balloon
- 19. Blanchard's Ascent, (Caricature)
- 20. Blanchard's Descent
- 21. Ascent from Dijon, 1784
- 22. Ascent of the Duke of Chartres
- 23. The "Minerva"
- 24. The First Attempt to Ascend in England
- 25. Blanchard
- 26. Dr. Jeffries
- 27. Coronation fete at Paris
- 28. The Wreck of the "Geant"
- 29. Pilatre des Roziers
- 30. Employment of a Balloon at the Battle of Fleurus

BALLOONS AND AIR JOURNEYS.

THE WRECK OF THE "GÉANT."

PART I. THE CONQUEST OF THE SKIES.— 1783.

Chapter I. Introduction.

The title of our introduction to aeronautics may appear ambitious to astronomers, and to those who know that the infinite space we call the heavens is for ever inaccessible to travellers from the earth; but it was not so considered by those who witnessed the ardent enthusiasm evoked at the ascension of the first balloon. No discovery, in the whole range of history, has elicited an equal degree of applause and admiration—never has the genius of man won a triumph which at first blush seemed more glorious. The mathematical and physical sciences had in aeronautics achieved apparently their greatest honours, and inaugurated a new era in the progress of knowledge. After having subjected the earth to their power; after having made the waves of the sea stoop in submission under the keels of their ships; after having caught the lightning of heaven and made it subservient to the ordinary purposes of life, the genius of man undertook to conquer the regions of the air. Imagination, intoxicated with past successes, could descry no limit to human power; the gates of the infinite seemed to be swinging back before man's advancing step, and the last was believed to be the greatest of his achievements.

In order to comprehend the frenzy of the enthusiasm which the first aeronautic triumphs called forth, it is necessary to recall the appearance of Montgolfier at Versailles, on the 19th of September, 1783, before Louis XVI, or of the earliest aeronauts at the Tuileries. Paris hailed the first of these men with the greatest acclaim, "and then, as now," says a French writer, "the voice of Paris gave the cue to France, and France to the world!" Nobles and artisans, scientific men and badauds, great and small, were moved with one universal impulse. In the streets the praises of the balloon were sung; in the libraries models of it abounded; and in the salons the one universal topic was the great "machine." In anticipation, the poet delighted himself with bird's-eye views of the scenery of strange countries; the prisoner mused on what might be a new way of escape; the physicist visited the laboratory in which the lightning and the meteors were manufactured; the geometrician beheld the plans of cities and the outlines of kingdoms; the general discovered the position of the enemy or rained shells on the besieged town; the police beheld a new mode in which to carry on the secret service; Hope heralded a new conquest from the domain of nature, and the historian registered a new chapter in the annals of human knowledge.

"Scientific discoveries in general," says Arago, "even those from which men expect the most advantage, like those of the compass and the steam-engine, were greeted at first with contempt, or at the best with indifference. Political events, and the fortunes of armies monopolised almost entirely the attention of the people. But to this rule there are two exceptions—the discoveries of America and of aerostatics, the advents of Columbus and of Montgolfier." It is not here our duty to inquire how it happened that the discoveries made by these two personages are classed together. Air-travelling may be as unproductive of actual good to society as filling the belly with the "east wind" is to the body, while every one knows something of the extent to which the discovery of Columbus has influenced the character, the civilisation, the destinies, in short, of the human race. We are speaking at present of the known and well-attested fact, that the discovery of America and the discovery of the method of traversing space by means of balloons—however they may differ in respect of results to man—rank equally in this, that of all other discoveries these two have attracted the greatest amount of attention, and given, in their respective eras, the greatest impulse to popular feeling. Let the reader recall the marks of enthusiasm which the discovery of the islands on the east coast of America excited in Andalusia, in Catalonia, in Aragon and Castile—let him read the narrative of the honours paid by town and village, not only to the hero of the enterprise, but even to his commonest sailors, and then let him search the records of the epoch for the degree of sensation produced by the discovery of aeronautics in France, which stands in the same relationship to this event as that in which Spain stands to the other. The processions of Seville and Barcelona are the exact prototypes of the fetes of Lyons and Paris. In France, in 1783, as in Spain two centuries previously, the popular imagination was so greatly excited by the deeds performed, that it began to believe in possibilities of the most unlikely description. In Spain, the conquestadores and their followers believed that in a few days after they had landed on American soil, they would have gathered as much gold and precious stones, as were then possessed by the richest European Sovereigns. In France, each one following his own notions, made out for himself special benefits to flow from the discovery of balloons. Every discovery then appeared to be only the precursor of other and greater discoveries, and nothing after that time seemed to be impossible to him who attempted the conquest of the atmosphere. This idea clothed itself in every form. The young embraced it with enthusiasm, the old made it the subject of endless regrets. When one of the first aeronautic ascents was made, the old Marechal Villeroi, an octogenarian and an invalid, was conducted to one of the windows of the Tuileries, almost by force, for he did not believe in balloons. The balloon, meanwhile, detached itself from its moorings; the physician Charles, seated in the car, gaily saluted the public, and was then majestically launched into space in his air-boat; and at once the old Marechal, beholding this, passed suddenly from unbelief to perfect faith in aerostatics and in the capacity of the human mind, fell on his knees, and, with his eyes bathed in tears, moaned out pitifully the words, "Yes, it is fixed! It is certain! They will find out the secret of avoiding death; but it will be after I am gone!"

If we recall the impressions which the first air-journeys made, we shall find that, among people of enthusiastic temperament, it was believed that it was not merely the blue sky above us, not merely the terrestrial atmosphere, but the vast spaces through which the worlds move, that were to become the domain of man—the sea of the balloon. The moon, the mysterious dwelling-place of men unknown, would no longer be an inaccessible place. Space no longer contained regions which man could not cross! Indeed, certain expeditions attempted the crossing of the heavens, and brought back news of the moon. The planets that revolve round the sun, the far-flying comets, the most distant stars—these formed the field which from that time was to lie open to the investigations of man.

This enthusiasm one can well enough understand. There is in the simple fact of an aerial ascent something so bold and so astonishing, that the human spirit cannot fail to be profoundly stirred by it. And if this is the feeling of men at the present day, when, after having been witnesses of ascents for the last eighty years, they see men confiding themselves in a swinging car into the immensities of space, what must have been the astonishment of those who, for the first time since the commencement of the world, beheld one of their fellow-creatures rolling in space, without any other assurance of safety than what his still dim perception of the laws of nature gave him?

Why should we be obliged here to state that the great discovery that stirred the spirits of men from the one end of Europe to the other, and gave rise to hopes of such vast discoveries, should have failed in realising the expectations which seemed so clearly justified by the first experiments? It is now eighty-six years since the first aerial journey astonished the world, and yet, in 1870, we are but little more advanced in the science than we were in 1783. Our age is the most renowned for its discoveries of any that the world has seen. Man is borne over the surface of the earth by steam; he is as familiar as the fish with the liquid element; he transmits his words instantaneously from London to New York; he draws pictures without pencil or brush, and has made the sun his slave. The air alone remains to him unsubdued. The proper management of balloons has not yet been discovered. More than that, it appears that balloons are unmanageable, and it is to air-vessels, constructed more nearly upon the model of birds, that we must go to find out the secret of aerial navigation. At present, as in former times, we are at the mercy of balloons—globes lighter than the air, and therefore the sport and the prey of tempests and currents. And aeronauts, instead of showing themselves now as the benefactors of mankind, exhibit themselves mainly to gratify a frivolous curiosity, or to crown with eclat a public fete.

Chapter II. Attempts in Ancient Times to Fly in the Air.

Before contemplating the sudden conquest of the aerial kingdom, as accomplished and proclaimed at the end of the last century, it is at once curious and instructive to cast a glance backward, and to examine, by the glimmering of ancient traditions, the attempts which have been made or imagined by man to enfranchise himself from the attraction of the earth.

"The greater number of the arts and sciences can be traced along a chronological ladder of great length: some, indeed, lose themselves in the night of time." The accomplishment of raising oneself in the air, however, had no actual professors in antiquity, and the discovery of Montgolfier seems to have come into the world, so to speak, spontaneously. By this it is to be understood that, unlike Copernicus and Columbus, Montgolfier could not read in history of any similar discovery, containing the germ of his own feat. At least, we have no proof that the ancient nations practiced the art of aerial navigation to any extent whatever. The attempts which we are about to cite do not strictly belong to the history of aerostatics.

Classic mythology tells us of Daedalus, who, escaping with his son Icarus from the anger of Minos, in the Isle of Crete, saved himself from the immediate evil by the aid of wings, which he made for himself and his son, and by means of which they were enabled to fly in the air. The wings, it appears, were soldered with wax, and Icarus, flying too high, was struck by a ray of the sun, which melted the wax. The youth fell into the sea, which from him derived its name of Icarian. It is possible that this fable only symbolisms the introduction of sails in navigation.

Coming down through ancient history, we note a certain Archytas, of Tarentum, who, in the fourth century B. C., is said to have launched into the air the first "flying stag," and who, according to the Greek writers, "made a pigeon of wood, which flew, but which could not raise itself again after having fallen." Its flight, it is said, "was accomplished by means of a mechanical contrivance, by the vibrations of which it was sustained in the air."

In the year 66 A.D., in the time of Nero, Simon, the magician—who called himself "the mechanician"—made certain experiments at Rome of flying at a certain height. In the eyes of the early Christians this power was attributed to the devil, and St. Peter, the namesake of this flying man, is said to have prayed fervently while Simon was amusing himself in space. It was possibly in answer to his prayers that the magician failed in his flight, fell upon the Forum, and broke his neck on the spot.

From the summit of the tower of the hippodrome at Constantinople, a certain Saracen met the same fate as Simon, in the reign of the Emperor Comnenus. His experiments were conducted on the principle of the inclined plane. He descended in an oblique course, using the resistance of the air as a support. His robe, very long and very large, and of which the flaps were extended on an osier frame, preserved him from suddenly falling.

The inclined plane probably suggested to Milton the flight of the angel Uriel, in "Paradise Lost," who descended in the morning from heaven to earth upon a ray of the sun, and ascended in the evening from earth to heaven by the same means. But we cannot quote here the fancies of pure imagination, and we will not speak of Medeus the magician, of the enchantress Armida, of the witches of the Brocken, of the hippogriff of Zephyrus with the rosy wings, or of the diabolical inventions of the middle ages, for many of which the stake was the only reward.

Roger Bacon, in the thirteenth century, inaugurated a more scientific era. In his "Treaty of the Admirable Power of Art and Nature," he puts forth the idea that it is possible "to make flying-machines in which the man, being seated or suspended in the middle, might turn some winch or crank, which would put in motion a suit of wings made to strike the air like those of a bird." In the same treatise he sketches a flying-machine, to which that of Blanchard, who lived in the eighteenth century, bears a certain resemblance. The monk, Roger Bacon, was worthy of entering the temple of fame before his great namesake the Lord Chancellor, who in the seventeenth century inaugurated the era of experimental science.

Jean Baptiste Dante, a mathematician of Perugia, who lived in the latter part of the fifteenth century, constructed artificial wings, by means of which, when applied to thin bodies, men might raise themselves off the ground into the air. It is recorded that on many occasions he experimented with his wings on the Lake Thrasymenus. These experiments, however, had a sad end. At a fete, given for the celebration of the marriage of Bartholomew d'Alvani, Dante, who must not be confounded with the poet, whose flights were of quite another kind—offered to exhibit the wonder of his wings to the people of Perugia. He managed to raise himself to a great height, and flew above the square; but the iron with which he moved one of his wings having been bent, he fell upon the church of the Virgin, and broke his thigh.

A similar accident befell a learned English Benedictine Oliver of Malmesbury. This ecclesiastic was considered gifted with the power of foretelling events; but, like other similarly circumstanced, he does not seem to have beer able to divine the fate which awaited himself. He constructed wings after the model of those which according to Ovid, Daedalus made use of. These he attached to his arms and his feet, and, thus furnished, he threw himself from the height of a tower. But the wings bore him up for little more than a distance of 120 paces. He fell at the foot of the tower, broke his legs, and from that moment led a languishing life. He consoled himself, however, in his misfortune by saying that his attempt must certainly have succeeded had he only provided himself with a tail.

Before going further, let us take notice that the seventeenth century is, par excellence, the century distinguished for narratives of imaginary travels. It was then that astronomy opened up its world of marvels. The knowledge of observers was vastly increased, and from that time it became possible to distinguish the surface of the moon and of other celestial bodies. Thus a new world, as it were, was revealed for human thought and speculation. We learned that our globe was not, as we had supposed, the centre of the universe. It was assigned its place far from that centre, and was known to be no more than a mere atom, lost amid an incalculable number of other globes. The revelations of the telescope proved that those who formerly were considered wise actually knew nothing. Quickly following these discoveries, extraordinary narratives of excursions through space began to be given to the world.

Those scientific romances were simply wild exaggerations, based upon the thinnest foundation of scientific facts. In order, however, to describe a journey among the stars, it was necessary to invent some mode of locomotion in these distant regions. In former times Lucian had been content with a ship which ascended to the rising moon upon a waterspout; but it was now necessary to improve upon this very primitive mode, as people began to know something more of the forces of nature. One of the first of these travellers in imagination to the moon in modern times was Godwin (1638), and his plan was more ingenious than that of Lucian. He trained a great number of the wild swans of St. Helena to fly constantly upward toward a white object, and, having succeeded in thus training them, one fine night he threw himself off the Peak of Teneriffe, poised upon a piece of board, which was borne upward to the white moon by a great team of the gigantic swans. At the end of twelve days he arrived, according to his story, at his destination. A little later another writer of this peculiar kind of fiction, Wilkins, an Englishman, professed to have made the same ascent, borne up by an eagle. Alexandre Dumas, who recently wrote a short romance upon the same subject, only made a translation of an English work by that author. Wilkins' work is entitled, "The Discovery of a New World." One chapter of the book bears the title, "That 'tis possible for some of our posterity to find out a conveyance to this other world; and, if there be inhabitants there, to have commerce with them." It is thus that the right reverend philosopher reasons:—

"If it be here inquired what means there may be conjectured for our ascending beyond the sphere of the earth's mathematical vigour, I answer.—1. 'Tis not possible that a man may be able to fly by the application of wings to his own body, as angels are pictured, as Mercury and Daedalus are feigned, and as hath been attempted by divers, particularly by a Turk in Constantinople, a Busbequius relates. 2. If there be such a great duck in Madagascar as Marcus Polus, the Venetian, mentions, the feathers of whose wings are twelve feet long, which can scoop up a horse and his rider, or an elephant, as our kites do a mouse; why, then, 'Tis but teaching one of these to carry a man, and he may ride up thither, as Ganymede does upon an eagle. 3. Or if neither of these ways will serve yet I do seriously, and upon good grounds, affirm it is possible to make a flying chariot, in which a man may sit and give such a motion to it as shall convey him through the air. And this, perhaps, might be made large enough to carry divers men at the same time, together with food for their viaticum, and commodities for traffic. It is not the bigness of anything in this kind that can hinder its motion if the motive faculty be answerable "hereunto. We see that; great ship swims as well as a small cork, and an eagle flies in the air as well as a little gnat. This engine may be contrived from the same principles by which Archytas made a wooden dove, and Regiomontanus a wooden eagle. I conceive it were no difficult matter (if a man had leisure) to show more particularly the means of composing it. The perfecting of such an invention would be of such excellent use that it were enough, not only to make a man famous but the age wherein he lives. For, besides the strange discoveries that it might occasion in this other world, it would be also of inconceivable advantage for travelling, above any other conveyance that is now in use. So that, notwithstanding all these seeming impossibilities, it is likely enough that there may be a means invented of journeying to the moon; and how happy shall they be that are first successful in this attempt!"

Afterwards comes Cyrano of Bergerac, who promulgates five different means of flying in the air. First, by means of phials filled with dew, which would attract and cause to mount up. Secondly, by a great bird made of wood, the wings of which should be kept in motion. Thirdly, by rockets, which, going off successively, would drive up the balloon by the force of projection. Fourthly, by an octahedron of glass, heated by the sun, and of which the lower part should be allowed to penetrate the dense cold air, which, pressing up against the rarefied hot air, would raise the balloon. Fifthly, by a car of iron and a ball of magnetised iron, which the aeronaut would keep throwing up in the air, and which would attract and draw up the balloon. The wiseacre who invented these modes of flying in the air seems, some would say, to have been more in want of very strict confinement on the earth than of the freedom of the skies.

In 1670 Francis Lana constructed the flying-machine shown on the next page. The specific lightness of heated air and of hydrogen gas not having yet been discovered, his only idea for making his globes rise was to take all the air out of them. But even supposing that the globes were thus rendered light enough to rise, they must inevitably have collapsed under the atmospheric pressure.

As for the idea of making use of a sail to direct the balloon, as one directs a vessel, that also was a delusion; for the whole machine, globes and sails, being freely thrown into the air, would infallibly follow the direction of the wind, whatever that might be. When a ship lies in the sea, and its sails are inflated with the wind, we must remember that there are two forces in operation—the active force of the wind and the passive force of the resistance of the water; and in working these forces the one against the other, the sailor can turn within a point of any direction he pleases. But when we are subjected wholly to a single force, and have no point of support by the use of which to turn that force to our own purposes, as is the case with the aeronaut, we are entirely at the mercy of that force, and must obey it.

After the flying-machine of Lana there was constructed by Galien (who, like the former, was an ecclesiastic) an air-boat, less chimerical in its form, looked at in view of the conditions of aerial navigation, but much more singular. Galien describes his air-boat, in 1755, in his little work entitled, "The Art of Sailing in the Air." His project was a most extraordinary one, and its boldness is only equalled by the seriousness of the narrative. According to him, the atmosphere is divided into two horizontal layers, the upper of which is much lighter than the lower. "But," says Galien, "a ship keeps its place in the water because it is full of air, and air is much lighter than water. Suppose, then, that there was the same difference of weight between the upper and the lower layer of air as there is between the lower stratum and water; and suppose, also, a boat which rested upon the lower layer of air, with its bulk in the lighter upper layer—like a ship which has its keel in the water but its bulk in the air—the same thing would happen with the air-ship as with the water-ship—it would float in the denser layer of air."

Galien adds that in the region of hail there was in the air a separation into two layers, the weights of which respectively are as 1 to 2. "Then," says he, "in placing an air-boat in the region of hail, with its sides rising eighty-three fathoms into the upper region, which is much more light, one could sail perfectly."

But how to get this enormous air-boat up to the region of hail? This is a minor detail, respecting which Galien is not clear.

From the labours of Lana and Galien, with their impossible flying machines, the inventor of the balloon could derive no benefit whatever; nor is his fame to be in the least diminished because many had laboured in the same field before him. Nor can the story of the ovoador, or flying man, a legend very confused, and of which there are many versions, have given to Montgolfier any valuable hints. It appears that a certain Laurent de Guzman, a monk of Rio Janeiro, performed at Lisbon before the king, John V., raising himself in a balloon to a considerable height. Other versions of the story give a different date, and assign the pretended ascent to 1709. The above engraving, extracted from the "Bibliotheque de la Rue de Richelieu," is an exact copy of Guzman's supposed balloon.

In 1678 a mechanician of Salle, in Maine, named Besnier invented a flying-machine. The machine consisted of four great wings, or paddles, mounted at the extremities of levers, which rested on the shoulders of the man who guided it, and who could move them alternately by means of his hands and feet. The following description of the machine is given in the Journal de Paris by an eye-witness:

"The 'wings' are oblong frames, covered with taffeta, and attached to the ends of two rods, adjusted on the shoulders The wings work up and down. Those in front are worked by the hands; those behind by the feet, which are connected with the ends of the rods by strings. The movements were such that when the right hand made the right wing descend in front, the left foot made the left wing descend behind; and in like manner the left hand in front and the right foot behind acted together simultaneously. This diagonal action appeared very well contrived; it was the action of most quadrupeds as well as of man when walking; but the contrivance, like others of the same kind, failed in not being fitted with gearing to enable the air traveller to proceed in any other direction than that in which the wind blew him. The inventor first flew down from a stool, then from a table, afterwards from a window, and finally from a garret, from which he passed above the houses in the neighbourhood, and then, moderating the working of his machine, he descended slowly to the earth."

Tradition records that under Louis XIV. a certain rope-dancer, named Alard, announced that on a certain day he would perform the feat of flying in the air. We have no description of his wings. It is recorded, however, that he set out on his adventurous flight; but he had not calculated all the necessities of the case, and, falling to the ground, he was dangerously hurt.

Leonardo da Vinci might have known the art of flying in the air, and might even have practiced it. A statement to this effect, at least, is found in several historians. We have, however, no direct proof of the fact.

The Abbe Deforges, of Etampes, announced in the journals in 1772 that he would perform the great feat. On the appointed day multitudes of the curious flocked to Etampes. The abbe's machine was a sort of gondola, seven feet long and about two feet deep. Gondola conductor, and baggage weighed in all 213 pounds. The pious man believed that he had provided against everything. Neither tempest nor rain should mar his flight, and there was no chance of his being upset; whilst the machine, he had decided, was to go at the rate of thirty leagues an hour.

The great day came, and the abbe, entering his air-boat amidst the applause of the spectators, began to work the wings with which it was provided with great rapidity. "But," says one who witnessed the feat, "the more he worked, the more his machine cleaved to the earth, as if it were part and parcel of it."

Retif de la Bretonne, in his work upon this subject, gives the accompanying picture of a flying man, furnished with very artistically designed wings, fitting exactly to the shoulders, and carrying a basket of provisions, suspended from his waist; and the frontispiece of the "Philosophic sans Pretention" is a view of a flying-machine. In the midst of a frame of light wood sits the operator, steadying himself with one hand, and with the other fuming a cremaillere, which appears to give a very quick rotatory movement to two glass globes revolving upon a vertical axis. The friction of the globes is supposed to develop electricity to which his power of ascending is ascribed.

To wings, however, aerial adventurers mostly adhered. The Marquis de Racqueville flew from a window of his hotel, on the banks of the Seine, and fell into a boat full of washerwomen on the river. All these unfortunate attempts were lampooned, burlesqued on the stage, and pursued with the mockery of the public.

Up to this time, therefore, the efforts of man to conquer the air had miscarried. They were conducted on a wrong principle, the machinery employed being heavier than the air itself But, even before the time of Montgolfier, the principles of aerostation began to be recognised, though nothing was actually done in the way of acting upon them. Thus, in 1767, Professor Black, of Edinburgh, announced in his class that a vessel, filled with hydrogen, would rise naturally in the air; but he never made the experiment, regarding the fact as capable of being employed only for amusement. Finally, Cavallo, in 1782, communicated to the Royal Society of London the experiments he had made, and which consisted in filling soap-bubbles with hydrogen. The bubbles rose in the atmosphere, the gas which filled them being lighter than air.

Chapter III. The Theory of Balloons.

A certain proposition in physics, known as the "Principle of Archimedes," runs to the following effect: — "Every body plunged into a liquid loses a portion of its weight equal to the weight of the fluid which it displaces." Everybody has verified this principle, and knows that objects are much lighter in water than out of it; a body plunged into water being acted upon by two forces — its own weight, which tends to sink it, and resistance from below, which tends to bear it up. But this principle applies to gas as well as to liquids — to air as well as to water. When we weigh a body in the air, we do not find its absolute weight, but that weight minus the weight of the air which the body displaces. In order to know the exact weight of an object, it would be necessary to weigh it in a vacuum.

If an object thrown into the air is heavier than the air which it displaces, it descends, and falls upon the earth; if it is of equal weight, it floats without rising or falling; if it is lighter, it rises until it comes to a stratum of air of less weight or density than itself. We all know, of course, that the higher you rise from the earth the density of the air diminishes. The stratum of air that lies upon the surface of the earth is the heaviest, because it supports the pressure of all the other strata that lie above. Thus the lightest strata are the highest.

The principle of the construction of balloons is, therefore, in perfect harmony with physical laws. Balloons are simply globes, made of a light, air-tight material, filled with hot air or hydrogen gas which rise in the air because (they are lighter than the air they displace).

The application of this principle appeared so simple, that at the time when the news of the invention of the balloon was spread abroad the astronomer Lalande wrote — "At this news we all cry, 'This must be! Why did we not think of it before?'" It had been thought of before, as we have seen in the last chapter, but it is often long after an idea is conceived that it is practically realised.

The first balloon, Montgolfier's, was simply filled with hot air; and it was because Montgolfier exclusively made use of hot air that balloons so filled were named Montgolfiers. Of course we see at a glance that hot air is lighter than cold air, because it has become expanded and occupies more space—that is to say, a volume of hot air contains actually less air than a volume of the same size of air that has not been heated. The difference between the weight of the hot air and the cold which it displaced was greater than the weight of tire covering of the balloon. Therefore the balloon mounted.

And, seeing that air diminishes in density the higher we ascend, the balloon can rise only to that stratum of air of the same density as the air it contains. As the warm air cools it gently descends. Again, as the atmosphere is always moving in currents more or less strong, the balloon follows the direction of the current of the stratum of air in which it finds itself.

Thus we see how simply the ascent of Montgolfiers, and their motions, are explained. It is the same with gas-balloons. A balloon, filled with hydrogen gas, displaces an equal volume of atmospheric air; but as the gas is much lighter than the air, it is pushed up by a force equal to the difference of the density of air and hydrogen gas. The balloon then rises in the atmosphere to where it reaches layers of air of a density exactly equal to its own, and when it gets there it remains poised in its place. In order that it may descend, it is necessary to let out a portion of the hydrogen gas, and admit an equal quantity of atmospheric air; and the balloon does not come to the ground till all, or nearly all, the gas has been expelled and common air taken in. Balloons inflated with hydrogen gas are almost the only ones in use at the present day. Scarcely ever is a Montgolfier sent up. There are aeronauts, however, who prefer a journey in a Montgolfier to one in a gas-balloon. The air voyager in this description of balloon had formerly many difficulties to contend with. The quantity of combustible material which he was bound to carry with him; the very little difference that there is between the density of heated and of cold air; the necessity of feeding the fire, and watching it without a moment's cessation, as it hangs in the rechaud over the middle of the car, rendered this sort of air travelling subject to many dangers and difficulties. Recently, M. Eugene Godard has obviated a portion of this difficulty by fitting a chimney, like that which is found of such incalculable service in the case of the Davy lamp. It is principally on account of this improvement that the Montgolfiere has risen so highly in popular esteem.

Generally it is not pure hydrogen that is made use of in the inflation of balloons. Aeronauts content themselves with the gas which we burn in our streets and houses, and thus it suffices, in inflating the balloon, to obtain from the nearest gas-works the quantity of gas necessary, and to lead it, by means of a pipe or tube, from the gasometer to the mouth or neck of the machine.

The balloon is made of long strips of silk, sewn together, and rendered air-tight by means of a coating of caoutchouc. A valve is fitted to the top, and by means of it the aeronaut can descend to the earth at will, by allowing some quantity of the gas to escape. The car in which he sits is suspended to the balloon by a network, which covers the whole structure. Sacks of sand are carried in this car as ballast, so that, when descending, if the aeronaut sees that he is likely to be precipitated into the sea or into a lake, he throws over the sand, and his air-carriage, being thus lightened, mounts again and travels away to a more desirable resting-place. The idea of the valve, as well as that of the sand ballast, is due to the physician Charles. They enable the aeronaut to ascend or descend with facility. When he wishes to mount, he throws over his ballast; when he wants to come down, he lets the gas escape by the valve at the roof of the balloon. This valve is worked by means of a spring, having a long rope attached to it, which hangs down through the neck to the car, where the aeronaut sits.

The operation of inflating a balloon with pure hydrogen is represented in the engraving on the next page.

Shavings of iron and zinc, water, and sulphuric acid, occupy a number of casks, which communicate, by means of tubes, with a central cask, which is open at the bottom, and is plunged in a copper full of water. The gas is produced by the action of the water and the sulphuric acid upon the zinc and the iron this is hydrogen mixed with sulphuric acid. In passing through the central copper, or vat, full of water, the gas throws off all impurities, and comes, unalloyed with any other matter, into the balloon by a long tube, leading from the central vats. In order to facilitate the entrance of the gas into the balloon two long poles are erected. These are furnished with pulleys, through which a rope, attached also to a ring at the top of the balloon, passes. By means of this contrivance the balloon can be at once lightly raised from the ground, and the gas tubes easily joined to it. When it is half full it is no longer necessary to suspend the balloon; on the contrary, it has to be secured, lest it should fly off. A number of men hold it back by ropes; but as the force of ascension is every moment increasing, the work of restraining the balloon is most difficult and exciting. At length, all preparations being complete, the car is suspended, the aeronaut takes his seat, the words "Let go all!" are shouted, and away goes the silken globe into space.

The balloon is never entirely filled, for the atmospheric pressure diminishing as it ascends, allows the hydrogen gas to dilate, in virtue of its expansive force, and, unless there is space for this expansion, the balloon is sure to explode in the air.

An ordinary balloon, with a lifting power sufficient to carry up three persons, with necessary ballast and materiel, is about fifty feet high, thirty-five feet in diameter' and 2,250 cubic feet in capacity. Of such a balloon, the accessories—the skin, the network, the car—would weigh about 335 lbs.

To find out the height at which he has arrived, the aeronaut consults his barometer. We know that it is the pressure of the air upon the cup of the barometer that raises the mercury in the tube. The heavier the air is, the higher is the barometer. At the level of the sea the column of mercury stands at 32 inches; at 3,250 feet—the air being at this elevation lighter—the mercury stands at 28 inches; at 6,500 feet above sea level it stands at 25 inches; at 10,000 feet it falls to 22 inches; at 20,000 feet to 15 inches. These, however, are merely the theoretic results, and are subject to some slight variation, according to locality, &c.

Sometimes the aeronaut makes his descent by means of the parachute, a separate and distinct contrivance. If, from any cause, it appears impracticable to effect a descent from the balloon itself, the parachute may be of the greatest service to the voyager at the present day it is chiefly used to astonish the public, by showing them the spectacle of a man who, from a great elevation in the air, precipitates himself into space, not to escape dangers which threaten him in his balloon, but simply to exhibit his courage and skill. Nevertheless, parachutes are often of great actual use, and aeronauts frequently attach them to their balloons as a precautionary measure before setting out on an aerial excursion.

The shape of a parachute, shown on the previous page, very much resembles that of the well-known all serviceable umbrella. The strips of silk of which it is formed are sewn together, and are bound at the top around a circular piece of wood. A number of cords, stretching away from this piece of wood, support the car in which the aeronaut is carried. At the summit is contrived an opening, which permits the air compressed by the rapidity of the descent to escape without causing damage to the parachute from the stress to which it is subjected.

The rapidity of the descent is arrested by the large surface which the parachute presents to the air. When the aeronaut wishes to descend by the parachute, all that is required is, after he has slipped down from the car of the balloon to that of the parachute, to loosen the rope which binds the latter to the former, which is done by means of a pulley. In an instant the aeronaut is launched into space with a rapidity in comparison with which the wild flights of the balloon are but gentle oscillations. But in a few moments, the air rushing into the folds of the parachute, forces them open like an umbrella, and immediately, owing to the wide surface which this contrivance presents to the atmosphere, the violence of the descent is arrested, and the aeronaut falls gently to the ground, without receiving too rude a shock.

The virtues of the parachute were first tried upon animals. Thus, Blanchard allowed his dog to fall in one from a height of 6,500 feet. A gust of wind caught the falling parachute, and swept it away up above the clouds. Afterwards, the aeronaut in his balloon fell in with the dog in the parachute, both of them high up in the cloudy reaches of the sky, and the poor animal manifested by his barking his joy at seeing his master. A new current separated the aerial voyagers, but the parachute, with its canine passenger, reached the ground safely a short time after Blanchard had landed from his balloon.

Experience has proved that, in the case of a descending parachute, if the rapidity of the descent is doubled the resistance of the air is quadrupled; if the rapidity is triple the resistance is increased ninefold; or, to speak in language of science, the resistance of the air is increased by the square of the swiftness of the body in motion. This resistance increases in proportion as the parachute spreads, and thus the uniformity of its fall is established a minute after it has been disengaged from the balloon. We can, therefore, check the descent of a body by giving it a surface capable of distension by the action of the air.

Garnerin, in the year 1802, conceived the bold design of letting himself fall from a height of 1,200 feet, and he accomplished the exploit before the Parisians. When he had reached the height he had fixed beforehand, he cut the rope which connected the parachute with the balloon. At first the fall was terribly rapid; but as soon as the parachute spread out the rapidity was considerably diminished. The machine made, however, enormous oscillations. The air, gathering end compressed under it, would sometimes escape by one side sometimes by the other, thus shaking and whirling the parachute about with a violence which, however great, had happily no unfortunate effect.

The origin of the parachute is more remote than is generally supposed, as there was a figure of one which appeared among a collection of machines at Venice, in 1617.

Another species of parachute, less perfect, to be sure; than that of Garnerin, but still a practical machine, was described 189 years before the great aeronaut's feat at Paris. We read in the narrative of the ambassador of Louis XIV at Siam, at the end of the seventeenth century, the following passage—"A mountebank at the court of the King of Siam climbed to the top of a high bamboo-tree, and threw himself into the air without any other support than two parasols. Thus equipped, he abandoned himself to the winds, which carried him, as by chance, sometimes to the earth, sometimes on trees or houses, and sometimes into the river, without any harm happening to him."

Is not this the idea of our parachutes?

Chapter IV. First Public Trial of the Balloon.

(Montgolfier's Balloon Annonay, 5th of June of 1783.)

We are accustomed to rank the brothers Joseph and Etienne Montgolfier as equally distinguished in the field of science. The reason for thus associating these two names seems to have been the fraternal friendship which subsisted in an extraordinary degree in the Montgolfier family, rather than any equality of claim which they had to the notice of posterity. After special investigation, we find that Joseph Montgolfier was very superior to his brother, and that it is to him principally, if not exclusively, that we owe the invention of aerostation. Nevertheless, we shall not insist upon this fact; and seeing that a sacred amity always cemented a perfect union in the Montgolfier family, we will regard that union as unbroken in any sense, and will not insinuate that the brother of Montgolfier was undeserving of the honoured rank which in his lifetime he held.

In 1783, the sons of Pierre Montgolfier, a rich papermaker at Annonay department of Ardeche, were already in the prime of life, and it is related of them that their principal occupation was experimenting in the physical sciences. Joseph Montgolfier, after being convinced by a number of minor experiments made in 1782 and 1783, that a heat of 180 degrees rarefied the air and made it occupy a space of TWICE the extent it occupied before being heated—or, in other words, that this degree of heat diminished the weight of air by one half—began to speculate on what might be the shape and the material of a structure which being filled with air thus heated, would be able to raise itself from the earth in spite of the weight of its own covering.

His first balloon was a small parallelopiped in very thin taffeta, containing less than seventy-eight cubic inches of air. He made it rise to the roof of his apartment in November, 1782—at Avignon, where he then happened to be. Having returned some little time after to Annonay, Joseph and his brother performed the same experiment, together in the open air with perfect success. Certain, then, of the new principle, they made a balloon of considerable size, containing upwards of sixty-five feet of heated air.

This machine likewise rose, tore away the cords by which it was at first held down, and mounting in the air to the height of from two to three hundred feet, fell upon the neighbouring hills after a considerable flight. The brothers Montgolfier then made a very large and strong balloon, with which they wished to bring their discovery before the public.

The appointed day was the 5th of June, 1783 and the nobility of the vicinity were invited to be present at the experiment. Faujas de Saint Fond, author of "La Description des Experiences de la Machine Aerostatique," published the same year, gives the following account of it:—

"What," says Saint Fond, "was the general astonishment when the inventors of the machine announced that immediately it should be full of gas, which they had the means of producing at will by the most simple process, it would raise itself to the clouds. It must be granted that, in spite of the confidence in the ingenuity and experience of the Montgolfiers, this feat seemed so incredible to those who came to witness it, that the persons who knew most about it—who were, at the same time, the most favourably predisposed in its favour—doubted of its success.

"At last the brothers Montgolfier commenced their work. They first of all began to make the smoke necessary for their experiment. The machine—which at first seemed only a covering of cloth, lined with paper, a sort of sack thirty-five feet high—became inflated, and grew large even under the eyes of the spectator, took consistence, assumed a beautiful form, stretched itself on all sides, and struggled to escape. Meanwhile, strong arms were holding it down until the signal was given, when it loosened itself, and with a rush rose to the height of 1,000 fathoms in less than ten minutes." It then described a horizontal line of 7,200 feet, and as it had lost a considerable amount of gas, it began to descend quietly. It reached the ground in safety; and this first attempt, crowned with such decisive success, secured for ever to the brothers Montgolfier the glory of one of the most astonishing discoveries.

"When we reflect for a moment upon the numberless difficulties which such a bold attempt entailed, upon the bitter criticism to which it would have exposed its projectors had it failed through any accident, and upon the sums that must have been spent in carrying it out, we cannot withhold the highest admiration for the men who conceived the idea and carried it out to such a successful issue."

Etienne Montgolfier has left us a description of this first balloon. "The aerostatic machine," he says, "was constructed of cloth lined with paper, fastened together on a network of strings fixed to the cloth. It was spherical; its circumference was 110 feet, and a wooden frame sixteen feet square held it fixed at the bottom. Its contents were about 22,000 cubic feet, and it accordingly displaced a volume of air weighing 1,980 1bs. The weight of the gas was nearly half the weight of the air, for it weighed 990 lbs., and the machine itself, with the frame, weighed 500: it was, therefore, impelled upwards with the force of 490 lbs. Two men sufficed to raise it and to fill it with gas, but it took eight to hold it down till the signal was given. The different pieces of the covering were fastened together with buttons and button-holes. It remained ten minutes in the air, but the loss of gas by the button-holes, and by other imperfections, did not permit it to continue longer. The wind at the moment of the ascent was from the north. The machine came down so lightly that no part of it was broken."

Chapter V. Second Experiment.

(Charles's Balloon, Paris, Champ de Mars, 27th of August, 1783.)

The indescribable enthusiasm caused by the ascent of the first balloon at Annonay, spread in all directions, and excited the wondering curiosity of the savants of the capital. An official report had been prepared, and sent to the Academy of Sciences in Paris, and the result was that the Academy named a commission of inquiry. But fame, more rapid than scientific commissions, and more enthusiastic than academies, had, at a single flight, passed from Annonay to Paris, and kindled the anxious ardour of the lovers of science in that city. The great desire was to rival Montgolfier, although neither the report nor the letters from Annonay had made mention of the kind of gas used by that experimenter to inflate his balloon. By one of the frequent coincidences in the history of the sciences, hydrogen gas had been discovered six years previously by the great English physician Cavendish, and it had hardly even been tested in the laboratories of the chemists when it all at once became famous. A young man well versed in physics, Professor Charles, assisted by two practical men, the brothers Robert, threw himself ardently into the investigation of the modes of inflating balloons with this gas, which was then called INFLAMMABLE AIR. Guessing that it was much lighter than that which Montgolfier had been obliged to make use of in his third-rate provincial town, Charles leagued himself with his two assistants to constrict a balloon of taffeta, twelve feet in diameter, covered with india-rubber, and to inflate it with hydrogen.

The thing thus arranged, a subscription was opened. The projected experiment having been talked of all over Paris, every one was struck with the idea, and subscriptions poured in. Even the most illustrious names are to be found in the list, which may be called the first national subscription in France. Nothing had been written of the forthcoming event in any public paper, yet all Paris seemed to flock to contribute to the curious experiment.

The inflation with hydrogen was effected in a very curious manner. As much as 1,125 lbs. of iron and 560 lbs. of sulphuric acid were found necessary to inflate a balloon which had scarcely a lifting power of 22 lbs., and the process of filling took no less than four hours. At length, however, at the end of the fourth hour, the balloon, composed of strips of silk, coated with varnish, floated, two-thirds full, from the workshop of the brothers Robert.

On the morning of the 26th of August, the day before the ascent was to be made, the balloon was visited at daybreak, and found to be in a promising state. At two o'clock on the following morning its constructors began to make preparations to transport it to the Champ de Mars, from which place it was to be let loose. Skilled workmen were employed in its removal, and every precaution was taken that the gas with which it was charged should not be allowed to escape. In the meantime the excitement of the people about this wonderful structure was rising to the highest pitch. The wagon on which it was placed for removal was surrounded on all sides by eager multitudes, and the night-patrols, both of horse and foot, which were set to guard the avenues leading to where it lay, were quite unable to stem the tide of human beings that poured along to get a glimpse of it.

The conveyance of the balloon to the Champ de Mars was a most singular spectacle. A vanguard, with lighted torches, preceded it; it was surrounded by special attendants, and was followed by detachments of night-patrols on foot and mounted. The size and shape of this structure, which was escorted with such pomp and precaution—the silence that prevailed—the unearthly hour, all helped to give an air of mystery to the proceedings. At last, having passed through the principal thoroughfares, it arrived at the Champ de Mars, where it was placed in an enclosure prepared for its reception.

When the dawn came, and the balloon had been fixed in its place by cords, attached around its middle and fixed to iron rings planted in the earth, the final process of inflation began.

The Champ de Mars was guarded by troops, and the avenues were also guarded on all sides. As the day wore on an immense crowd covered the open space, and every advantageous spot in the neighborhood was crowded with people. At five o'clock the report of a cannon announced to the multitudes, and to scientific men who were posted on elevations to make observations of the great event, that the grand moment had come. The cords were withdrawn, and, to the vast delight and wonder of the crowd assembled, the balloon shot up with such rapidity that in two minutes it had ascended 488 fathoms. At this height it was lost in a cloud for an instant, and, reappearing, rose to a great height, and was again lost in higher clouds. The ascent was a splendid success. The rain that fell damped neither the balloon nor the ardor of the spectators.

This balloon was 12 feet in diameter, 38 feet in circumference, and had a capacity of 943 cubic feet. The weight of the materials of which it was constructed was 25 lbs., and the force of ascension was that of 35 lbs.

The fall of the balloon was caused by the expansion and consequent explosion of the hydrogen gas. This event took place some distance out in the country, close to a number of peasants, whose terror at the sight and the sound of this strange monster from the skies was beyond description. The people assembled, and two monks having told them that the burst balloon was the hide of a monstrous animal, they immediately began to assail it vigorously with stones, flails, and pitchforks. The cure of the parish was obliged to walk up to the balloon to reassure his terrified flock. They finally attached the burst envelope to a horse's tail, and dragged it far across the fields.

Many drawings and engravings of the period represent the peasants armed with pitchforks, flails, and scythes, assailing it, a dog snapping at it, a garde-champetre firing at it, a fat priest preaching at it, and a troop of young people throwing stones at the unfortunate machine.

The news of this fiasco came to Paris, but too late. When search was made for the covering, scarcely a fragment could be found.

A somewhat humorous result of all this was the issue of a communication from government to the people, entitled, "Warning to the People on kidnapping Air-balloons." This document, duly signed and approved of, describes the ascents at Annonay and at Paris, explains the nature and the causes of the phenomena, and warns the people not to be alarmed when they see something like a "black moon" in the sky, nor to give way to fear, as the seeming monster is nothing more than a bag of silk filled with gas.

This first ascent in Paris was an important event. Every one, from the smallest to the greatest, was deeply interested in it, while to the man of science it was one of the most exciting of incidents. For the purpose of observing the altitude to which the balloon rose, and the course it took, Le Gentil was on the observatory, Prevost was on one of the towers of Notre Dame, Jeaurat was on La Place Louis XV., and d'Agelet was on the Champ de Mars. It was only Lalande that frowned as he witnessed the success of the experiment. He had predicted the year before that air-navigation was impossible.

Chapter VI. Third Experiment.

(Montgolfier's Balloon, Paris, Faubourg St. Antoine.)

As we have seen, the triumph of aerostation was sudden and complete. The young Montgolfier had arrived in Paris prior to the experiment of the 27th of August, and was present as a simple spectator on that occasion. immediately afterwards he set to work upon a balloon, which was to be made use of when the Academy should investigate the phenomenon at Versailles in presence of the king, Louis XVI.

It was at this time (September, 1783) that those small balloons, made of gold-beaters' skin, which are used as children's toys to the present day, were first made. The whole of Paris amused itself with them, repeating in little the phenomenon of the great ascent. The sky of the capital found itself all at once traversed by a multitude of small rosy clouds, formed by the hand of man.

Faujas de Saint Fond says that at first an attempt was made to construct balloons of fine, light paper; but this material being permeable, and the gas being inflammable, balloons thus made did not succeed. It was necessary to seek a material less porous, and, if possible, still lighter.

The Journal de Paris, of the 11th of September, 1783, informed the public that the Baron de Beaumanoir, "who cultivated the sciences and the fine arts with as much success as zeal," would send up a balloon eighteen inches in diameter. At noon of the same day he made this experiment in presence of a numerous assembly in the garden in front of the Hotel de Surgeres.. The little balloon mounted freely, but was held in, like a kite, by means of a silk thread. In the course of the same afternoon, the baron took down the balloon and filled it anew with hydrogen, and then let it off. The spectators had the pleasure of seeing it rise to a great height, and pass away in the direction of Neuilly, and it is said to have been found at a distance of several leagues, by peasants.

However trifling this experiment may appear at first sight, it added a new fact to the science of aerostation. The material employed by the baron was lighter and better than paper. It was what is called gold-beaters' skin. This skin is simply the interior lining of the large bowel of the ox. It is carefully prepared, is relieved of the fat, stringy and uneven parts, is dried, and is afterwards softened. Little balloons of this material came to be the fashion, and they are still frequently seen.

At the same time, Montgolfier was busy constructing, at the request of the Academy of Sciences, a balloon seventy feet high and forty in diameter, with which it was proposed to repeat the experiment of Annonay. He took up his quarters in the magnificent gardens of his friend Reveillon, proprietor of the royal manufactory of stained paper in the Faubourg St. Antoine. The new balloon was of a very singular shape: the upper part represented a prism, twenty-four feet high the top was a pyramid of the same height; the lower part was a truncated cone, twenty feet in depth. It was made of packing-cloth, lined with good paper, both inside and out.

The gossipping and prolix Faujas de Saint Fond thus describes this machine: — "It was painted blue, represented a sort of tent, and was richly ornamented with gold Its height was seventy feet; its weight 1,000 lbs.; the air which it displaced was 4,500 lbs. in volume, and the vapor with which it was filled was half the weight of ordinary air. The approach of the equinox having brought rain, all the conditions under which this balloon was constructed and exhibited were unfavourable. The structure was so large that it was impossible to get it together and stitch it, except in the open air — in the garden, in fact, where Montgolfier commenced its construction. It was a great labour to turn and fold this heavy covering, while the liability of the thick paper to crack was an additional difficulty. Not less than twenty men were required to move it, and they were obliged to use all their skill, and every precaution, not to destroy it. No balloon had ever given so much trouble. On the 11th of September the weather improved, and the balloon was entirely completed and prepared for the first experiment. In the evening the attempt was made. It was with admiration that the beholders saw the beautiful machine filling itself in the short space of nine minutes, swelling out on all sides and showing the full symmetry of its artistic form. It was firmly held in hand, or it would have risen to a great height. On the following day the actual ascent was to take place, and the commissioners of the Academy of Sciences were invited to be present. In the morning thick clouds covered the horizon, and a tempest was expected; but as there was an ardent desire that the ascent should take place without delay, and as all the gearing was in order, it was resolved to proceed.

"Fifty pounds of dry straw were fired in parcels under the balloon, and upon the fire were thrown at intervals several pounds of wool. This fuel produced in ten minutes such a volume of smoke that the huge balloon was speedily filled. It rose, with a weight of 500 lbs. holding it down, to some height above the ground, and had the ropes by which it was attached to the ground been cut, it would have mounted to a great height. Meantime the storm broke, rain descended, and the wind blew with great force. The most likely means of saving the balloon was to let it fly but as it was to ascend again on another occasion, at Versailles, the greatest efforts were made to bring it down, and these, together with the damage caused by the storm, eventually rent it into numberless fragments and tatters. It withstood the storm for twenty-four hours; then, however, the paper came peeling off, and this beautiful structure was a wreck."

Chapter VII. Fourth Experiment.

(Versailles, 19th September, 1783, in presence of Louis of XVI.)

Of course another balloon was wanted for the fete at Versailles. The king had demanded an ascent for the 19th, a week after the disaster at the Faubourg St. Antoine. Already the possibility of a man going up with the balloon was discussed, and people indulged in visions of splendid aerial trips; but the king would not hear of the proposal. Balloons were novelties, not offering sufficient security, and he was unwilling that any of his subjects should risk their lives in attempting the unknown. He consented, however, to a proposal that animals might be sent up in the first instance, by way of experiment, suspended in an osier cage attached to the neck of the balloon.

Montgolfier at once began a new balloon. A few days only were at his disposal; but, assisted by friends, he worked with such ardour and success that he was able, on the date appointed, to produce a magnificent spherical balloon, much stronger than the former, constructed of good strong cotton cloth, and painted in distemper.

It is proper here to remark that the first balloons were much more elegant in appearance than those afterwards made. The coloured prints and engravings of the period enable us to form an opinion of the splendour of their ornamentation and the beauty of their design. Sometimes the figures painted upon them represented scenes from the heathen mythology, and sometimes historical scenes; while rich embroideries, royal insignia, and gaily-coloured draperies added much to the general effect. The Versailles balloon was painted blue, with ornaments of gold, and it presented the form of a richly decorated tent. It was fifty-seven feet in height, and sixty-seven in diameter.

It was first tried at Paris, and succeeded perfectly. On the morning of the 19th it was carried to Versailles, where due preparation had been made for its reception In the great court of the castle a sort of theatre had been temporarily erected with a scaffolding, covered throughout with tapestry In the middle was an opening more than fifteen feet in diameter, in which was spread a banquet for those who had constructed the balloon. A numerous guard formed a double cordon around the structure. A raised platform was used for the fire by means of which the balloon was to be inflated; a covered funnel or chimney of strong cloth, painted, was suspended over the fire-place, and received the hot smoke as it arose. Through this funnel the heated air ascended straight up into the balloon.

At six in the morning, the road from Paris to Versailles was covered with carriages. Crowds came from all parts, and at noon the avenues, the square of the castle, the windows, and even the roofs of the houses, were crowded with spectators. The noblest, the most illustrious, and most learned men in France were present, and the splendour of the scene was complete when their majesties and the royal family entered within the enclosure, and went forward to inspect the balloon, and to make themselves familiar with the preparations for the ascent.

In a short time the fire was lit, the funnel extended over it, and the smoke rose inside, while the balloon, unfolding, gradually swelled to its full size, and then, drawing after it the cage, in which a sheep and some pigeons were enclosed, rose majestically into the air. Without interreruption, it ascended to a vast height, where, inclining toward the north, it seemed to remain stationary for a few seconds, showing all the beauty of its form, and then, as though possessed of life, it descended gently upon the wood of Vaucresson, 10,200 feet from the point of its departure. Its highest elevation, as estimated by the astronomers Le Gentil and M. Jeaurat, Jeaurat, was about 1,700 feet.

Chapter VIII. Men and Balloons.

It is not natural that the human mind should stop upon the way to the solution of a problem, especially when it seems to be on the point of arriving at a satisfactory conclusion to its labours. The osier cage of Versailles very soon transformed itself into a car, bearing human passengers, and the age of the "Thousand and One Nights" was expected to come back again. It was resolved to continue experiments, with the direct object of finding out whether it was impossible or desperately dangerous for man to travel in balloons. Montgolfier returned from Versailles, and constructed a new machine in the gardens of the Faubourg St. Antoine. It was completed on the 10th of October Its form was oval, its height 70 feet, its diameter 46 feet and its capacity 60,000 cubic feet. The upper part, embroidered with fleurs-de-lis, was further ornamented with the twelve signs of the zodiac, worked in gold. The middle part bore the monogram of the king, alternating with figures of the sun, while the lower part was garnished with masks, garlands, and spread eagles. A circular gallery made of osiers and festooned with draperies and other ornaments, was attached by a set of cords to the bottom of the structure. The gallery was three feet wide, and was protected by a parapet over three feet in height. It did not in any way interfere with the opening at the neck of the balloon, under which was suspended a grating of iron wire upon which the occupants of the gallery, who were to be provided with dried straw and wool, could in a few minutes kindle a fire and create fresh smoke, when that in the balloon began to be exhausted. The machine weighed, in all, 1,600 lbs. The public had previously been warned, in the Journal de Paris de Paris, that the approaching experiments were to be of a strictly scientific character; and as they would be only interesting to savants, they would not afford amusement for the merely curious. This announcement was necessary, to abate in some degree the excitement of the people until some satisfactory results should be obtained; it was also necessary for those engaged in the work, whose firmness of nerve might have suffered from the enthusiastic cries of excited spectators. On Wednesday, the 15th of October, Pilatre des Roziers, who had on other occasions given proofs of his intelligence and courage in performing dangerous feats, and who had already signalised himself in connection with balloons, offered to go up in the new machine. His offer was accepted; the balloon was inflated; stout ropes, more than eighty feet long, were attached to it, and it rose from the ground to the height to which this tackle allowed it. At this elevation it remained four minutes twenty-five seconds; and it is not surprising to hear that Roziers suffered no inconvenience from the ascent. What was really the interesting thing in this experiment was, that it showed how a balloon would fall when the hot air became exhausted, this being the point which caused the greatest amount of disquietude among men of science. In this instance the balloon fell gently; its form distended at the same time, and, after touching the ground, it rose again a foot or two, when its human passenger had jumped out.

On Friday, the 17th of October, this experiment was repeated, and the excitement of the public on this occasion was unbounded. "All the world" came to see. Roziers was again lifted up in the balloon, to the height of eighty feet; but so strong was the wind, and the strain on the ropes was so great, that the balloon was somewhat unsteady, and the exhibition was not on the whole such a splendid success as that of the preceding Wednesday.

On Sunday, Montgolfier chose a fine day for the following ascents: — "First Ascent: On the 19th of October, 1783, at half-past four, in presence of two thousand spectators, 'the machine' was filled with gas in five minutes, and Roziers, being placed in the gallery with a counterbalancing weight of 110 lbs. in the other side of the gallery, was carried up to the height of 200 feet. The machine remained six minutes at this elevation without any fire in the grating. Second Ascent: The machine carried Roziers and the counterbalancing weight—fire being in the grating—to the height of 700 feet. At this height it remained stationary eight and a half minutes As it was drawn back, a wind from the east bore it against a tuft of very tall trees in a neighbouring garden, where it got entangled, without, however, losing its equilibrium. The gas was renewed by Roziers, and the balloon again rising, extricated itself from among the branches, and soared majestically into the air, followed by the acclamations of the public. This second ascent was very instructive, for it had been often asserted that if ever a balloon fell upon a forest it would be destroyed, and would place those who travelled in it in the greatest peril. This experiment proved that the balloon does not FALL it DESCENDS; that it does not overturn; that it does not destroy itself on trees; that it neither causes death, nor even damage, to its passengers; that, on the contrary, the latter, by making new gas, give it the power of detaching itself from the trees; and that it can resume its course after such an event. The intrepid Roziers gave in this ascent a further proof of the facility he had in descending and ascending at will. When the machine had risen to the height of 200 feet it began to descend lightly, and just before it came to the earth the aeronaut very cleverly and quickly threw on more fuel and produced more smoke, at which the balloon, to the astonishment of every one, suddenly soared away again to its former elevation. Third Ascent: The balloon rose again with Roziers, accompanied this time by another aeronaut, Gerond de Villette; and as the cords had been lengthened, the adventurers were carried up to the height of 324 feet. At this elevation the balloon rested in perfect equilibrium for nine minutes. It was the first time that human beings had ever been carried to an equal elevation, and the spectators were astonished to find that they could remain there without danger and without alarm. The balloon had a superb effect at this elevation; it looked down upon the whole town, and was seen from all the suburbs. Its size seemed hardly diminished in the least, though the men themselves were barely visible. By the aid of glasses, Roziers could be seen calmly and industriously making new gas. When the balloon descended the two men declared that they had not experienced the slightest inconvenience from the elevation. They received the universal applause which their zeal and courage so well deserved. The Marquis d'Arlandes, a major of infantry, afterwards went up with Roziers, and this latter experiment was as successful as the former."

Some days after these experiments the conductors of the Journal de Paris who described them, received a letter from Montgolfier, and also one from Gerond de Villette. The latter only is of interest here. Gerond de Villette says: "I found myself in the space of a quarter of a minute raised 400 feet above the surface of the earth. Here we remained six minutes. My first employment was to watch with admiration my intelligent companion. His intelligence, his courage and agility in attending to the fire, enchanted me. Turning round, I could behold the Boulevards, from the gate of St. Antoine to that of St. Martin, all covered with people, who seemed to me a flat band of flowers of various colours. Glancing at the distance, I beheld the summit of Montmartre, which seemed to me much below our level. I could easily distinguish Neuilly, St. Cloud, Sevres, Issy, Ivry, Charenton, and Choisy. At once I was convinced that this machine, though a somewhat expensive one, might be very useful in war to enable one to discover the position of the enemy, his manoeuvres, and his marches; and to announce these by signals to one's own army. 1 believe that at sea it is equally possible to make use of this machine. These prove the usefulness of the balloon, which time will perfect for us. All that I regret is that I did not provide myself with a telescope."

Chapter IX. The First Aerial Voyage—Roziers and Arlandes.

> These experiments had only one aim—the application of Montgolfier's discovery to aerial navigation. The knowledge gained in the Faubourg St. Antoine having led to the most favourable conclusions, it was resolved that a first aerial voyage should be attempted.

"If," says Linguet, "there existed an autograph journal, written by Columbus, descriptive of his first great voyage with what jealous care it would be preserved, with what confidence it would be quoted! We should delight to follow the candid account which he gave of his thoughts, his hopes, his fears; of the complaints of his followers, of his attempts to calm them, and, finally, of his joy in the moment which, ratifying his word and justifying his boldness, declared him the discoverer of a new world All these details have been transmitted to us, but by stranger hands; and, however interesting they may be, one cannot help feeling that this circumstance makes them lose part of their value."

The narrative of the first aerial voyage, written by one of the two first aeronauts, exists, and we are in a position to place it before our readers. Such an enterprise certainly demanded great courage in him who was the first to dare to confide himself to the unknown currents of the atmosphere It threatened him with dangers, perhaps with death by a fill, by fire, by cold, or by straying into the mysterious cloud-land. Two men opposed the first attempt. Montgolfier temporised, the king forbade it, or rather only gave his permission on the condition that two condemned criminals should be placed in the balloon! "What!" cried Roziers, in indignation at the king's proposal, "allow two vile criminals to have the first glory of rising into the sky! No, no; that will never do!" Roziers conjured, supplicated, agitated in a hundred ways for permission to try the first voyage. He moved the town and the court; he addressed himself to those who were most in favour at Versailles; he pleaded with the Duchess de Polignac, who was all-powerful with the king. She warmly supported his cause before Louis. Roziers dispatched the Marquis d'Arlandes, who had been up with him, to the king. Arlandes asserted that there was no danger, and, as proof of his conviction, he offered himself to accompany Roziers. Solicited on all sides, Louis at last yielded.

The gardens of La Muette, near Paris, were fixed upon as the spot from which this aerial expedition should start. The Dauphin and his suite were present on the occasion. It was on the 21st of October, 1783, at one o'clock p.m., that Roziers and Irelands took their leave of the earth for the first time. The following is Arlandes' narrative of the expedition, given in the form of a letter, addressed by the marquis to Faujas de Saint Fond: — "You wish, my dear Faujas, and I consent most willingly to your desires, that, owing to the number of questions continually addressed to me, and for other reasons, I should gratify public curiosity and fix public opinion upon the subject of our aerial voyage.

"I wish to describe as well as I can the first journey which men have attempted through an element which, prior to the discovery of MM. Montgolfier, seemed so little fitted to support them.

"We went up on the 21st of October, 1783, at near two o'clock, M. Roziers on the west side of the balloon, I on the east. The wind was nearly north-west. The machine, say the public, rose with majesty; but really the position of the balloon altered so that M. Roziers was in the advance of our position, I in the rear.

"I was surprised at the silence and the absence of movement which our departure caused among the spectators, and believed them to be astonished and perhaps awed at the strange spectacle; they might well have reassured themselves I was still gazing, when M. Roziers cried to me —

"'You are doing nothing, and the balloon is scarcely rising a fathom.'

"'Pardon me,' I answered, as I placed a bundle of straw upon the fire and slightly stirred it. Then I turned quickly, but already we had passed out of sight of La Muette. Astonished, I cast a glance towards the river. I perceived the confluence of the Oise. And naming the principal bends of the river by the places nearest them, I cried, 'Passy, St. Germain, St. Denis, Sevres!'

"'If you look at the river in that fashion you will be likely to bathe in it soon,' cried Roziers. 'Some fire, my dear friend, some fire!'

"We travelled on; but instead of crossing the river, as our direction seemed to indicate, we bore towards the Invalides, then returned upon the principal bed of the river, and travelled to above the barrier of La Conference, thus dodging about the river, but not crossing it.

"'That river is very difficult to cross,' I remarked to my companion.

"'So it seems,' he answered; 'but you are doing nothing I suppose it is because you are braver than I, and don't fear a tumble.'

"I stirred the fire, I seized a truss of straw with my fork; I raised it and threw it in the midst of the flames. An instant afterwards I felt myself lifted as it were into the heavens.

"'For once we move,' said I.

"'Yes, we move,' answered my companion.

"At the same instant I heard from the top of the balloon a sound which made me believe that it had burst. I watched, yet I saw nothing. My companion had gone into the interior, no doubt to make some observations. As my eyes were fixed on the top of the machine I experienced a shock, and it was the only one I had yet felt. The direction of the movement was from above downwards I then said —

"'What are you doing? Are you having a dance to yourself?'

"'I'm not moving.'

"'So much the better. It is only a new current which I hope will carry us from the river,' I answered.

"I turned to see where we were, and found we were between the Ecole Militaire and the Invalides.

"'We are getting on.' said Roziers.

"'Yes, we are travelling.'

"'Let us work, let us work,' said he.

"I now heard another report in the machine, which I believed was produced by the cracking of a cord. This new intimation made me carefully examine the inside of our habitation. I saw that the part that was turned towards the south was full of holes, of which some were of a considerable size.

"'It must descend,' I then cried.

"'Why?'

"'Look!' I said. At the same time I took my sponge and quietly extinguished the little fire that was burning some of the holes within my reach; but at the same moment I perceived that the bottom of the cloth was coming away from the circle which surrounded it.

"'We must descend,' I repeated to my companion.

"He looked below.

"'We are upon Paris,' he said.

"'It does not matter,' I answered 'Only look! Is there no danger? Are you holding on well?'

"'Yes.'

"I examined from my side, and saw that we had nothing to fear. I then tried with my sponge the ropes which were within my reach. All of them held firm. Only two of the cords had broken.

"I then said, 'We can cross Paris.'

"During this operation we were rapidly getting down to the roofs. We made more fire, and rose again with the greatest ease. I looked down, and it seemed to me we were going towards the towers of St. Sulpice; but, on rising, a new current made us quit this direction and bear more to the south. I looked to the left, and beheld a wood, which I believed to be that of Luxembourg. We were traversing the boulevard, and I cried all at once—

"'Get to ground!'

"But the intrepid Roziers, who never lost his head, and who judged more surely than I, prevented me from attempting to descend. I then threw a bundle of straw on the fire. We rose again, and another current bore us to were now close to the ground, between two mills. As soon to the left. We as we came near the earth I raised myself over the gallery, and leaning there with my two hands, I felt the balloon pressing softly against my head. I pushed it back, and leaped down to the ground. Looking round and expecting to see the balloon still distended, I was astonished to find it quite empty and flattened. On looking for Roziers I saw him in his shirt-sleeves creeping out from under the mass of canvas that had fallen over him. Before attempting to descend he had put off his coat and placed it in the basket. After a deal of trouble we were at last all right.

"As Roziers was without a coat I besought him to go to the nearest house. On his way thither he encountered the Duke of Chartres, who had followed us, as we saw, very closely, for I had had the honour of conversing with him the moment before we set out."

The following report of this first aerial voyage was drawn up by scientific observers, among other signatures to it being that of Benjamin Franklin.

"Today 21st of October, 1783, at the Chateau de la Muette, an experiment was made with the aerostatic machine of M. Montgolfier. The sky was clouded in many parts, clear in others—the wind north-west. At mid-day a signal was given, which announced that the balloon was being filled. Soon after, in spite of the wind, it was inflated in all its parts, and the ascent was made. The Marquis d'Arlandes and M. Pilatre des Roziers were in the gallery. The first intention was to raise the machine and pull it back with ropes, to test it, to find out the exact weight which it could carry, and to see if everything was properly arranged before the actual ascent was attempted. But the machine, driven by the wind, far from rising vertically, was directed upon one of the walks of a garden, and the cords which held it shook with so much force that several rents were made in the balloon. The machine, being brought back to its place, was repaired in less than two hours. Being again inflated, it rose once more, bearing the same persons, and when it had risen to the height of 250 feet, the intrepid voyagers, bowing their heads, saluted the spectators. One could not resist a feeling of mingled fear and admiration. Soon the aeronauts were lost to view, but the balloon itself, displaying its very beautiful shape, mounted to the height of 3,000 feet, and still remained visible. The voyagers, satisfied with their experience, and not wishing to make a longer course, agreed to descend, but, perceiving that the wind was driving them upon the houses of the Rue de Sevres, preserved their self-possession, renewed the hot air, rose anew and continued their course till they had passed Paris.

"They then descended tranquilly in the country, beyond the new boulevard, without having experienced the slightest inconvenience, having still the greater part of their fuel untouched. They could, had they desired, have cleared a distance three times as great as that which they traversed. Their flight was nearly 30,000 feet, and the time it occupied was from twenty to twenty-five minutes. This machine was 70 feet high, 46 feet in diameter, and had a capacity of 60,000 cubic feet."

It is reported that Franklin, more illustrious in his humility than the most brilliant among the lords of the court, when consulted respecting the possible use of balloons, answered simply, "C'est l'enfant qui vient de naitre?"

Chapter X. The Second Arial Voyage.

(1st December 1783. — Charles and Robert at the Tuileries.)

The first ascent of Roziers and Arlandes was a feat of hardihood almost unique. The men's courage was, so to speak, their only guarantee. Thanks to the balloon, however, they accomplished one of the most extraordinary enterprises ever achieved by our race.

On the day after the experiment of the Champ de Mars (27th of August), Professor Charles — who had already acquired celebrity at the Louvre, by his scientific collection and by his rank as an official instructor — and the Brothers Robert, mechanicians, were engaged in the construction of a balloon, to be inflated with hydrogen gas, and destined to carry a car and one or two passengers. For this ascent Charles may be said to have created all at once the art of aerostation as now practiced, for he brought it at one bound to such perfection that since his day scarcely any advance has been made upon his arrangements. His simple yet complete invention was that of the valve which gives escape to the hydrogen gas, and thus renders the descent of the balloon gentle and gradual; the car that carries the travellers; the ballast of sand, by which the ascent is regulated and the fall is moderated; the coating of caoutchouc, by means of which the material of the balloon is rendered airtight and prevents loss of gas; and, finally, the use of the barometer, which marks at every instant, by the elevation or the depression of the mercury, the position in which the aeronaut finds himself in the atmosphere. Charles created all the contrivances, or, in other words, all the ingenious precautions which make up the art of aerostation.

On the 26th of November, the balloon, fitted with its network, and having the car attached to it, was sent away from the hall of the Tuileries, where it had been exhibited. The ascent was fixed for the 1st of December, 1783, a memorable day for the Parisians.

At noon upon that day, the subscribers, who had paid four louis for their seats, took their places within the enclosure outside the circle, in which stood the casks employed for making the gas. The humbler subscribers, at three francs a-head, occupied the rest of the garden. The number of spectators, as we read underneath the numerous coloured prints which represent this spectacle, was 600,000; but though, without doubt, the gardens of the Tuileries are very large, it is probable this figure is a considerable overstatement, for this number would have been three-fourths of the whole population of Paris.

The roofs and windows of the houses were crowded, whilst the Pont Royal and the square of Louis XV. were covered by an immense multitude. About mid-day a rumour was spread to the effect that the king forbade the ascent. Charles ran to the Chief Minister of State, and plainly told him that his life was the king's, but his honour was his own: his word was pledged to the country and he would ascend. Taking this high ground, the bold professor gained an unwilling permission to carry out his undertaking.

A little afterwards the sound of cannon was heard. This was the signal which announced the last arrangements and thus dissipated all doubt as to the rising of the balloon, There had during the day been considerable disturbance among the crowd, between the partisans of Charles and Montgolfier; each party extolled its hero, and did everything possible to detract from the merits of the rival inventor. But whatever ill-feeling might have existed was swept away by Professor Charles with a compliment. When he was ready to ascend, he walked up to Montgolfier, and, with the true instinct of French politeness, presented him with a little balloon, saying at the same time—

"It is for you, monsieur, to show us the way to the skies."

The exquisite taste and delicacy of this incident touched the bystanders as with an electric shock, and the place at once rang out with the most genuine and hearty applause The little balloon thrown up by Montgolfier sped away to the north-east, its beautiful emerald colour showing to fine effect in the sun.

From this point let us follow the narrative of Professor Charles himself.

"The balloon," he says, "which escaped from the hands of M. Montgolfier, rose into the air, and seemed to carry with it the testimony of friendship and regard between that gentleman and myself, while acclamations followed it. Meanwhile, we hastily prepared for departure. The stormy weather did not permit us to have at our command all the arrangements which we had contemplated the previous evening; to do so would have detained us too long upon the earth. After the balloon and the car were in equilibrium, we threw over 19 lbs. of ballast, and we rose in the midst of silence, arising from the emotion and surprise felt on all sides.

"Nothing will ever equal that moment of joyous excitement which filled my whole being when I felt myself flying away from the earth. It was not mere pleasure; it was perfect bliss. Escaped from the frightful torments of persecution and of calumny, I felt that I was answering all in rising above all.

"To this sentiment succeeded one more lively still—the admiration of the majestic spectacle that spread itself out before us. On whatever side we looked, all was glorious; a cloudless sky above, a most delicious view around. 'Oh, my friend,' said I to M. Robert, 'how great is our good fortune! I care not what may be the condition of the earth; it is the sky that is for me now. What serenity! what a ravishing scene! Would that I could bring here the last of our detractors, and say to the wretch, Behold what you would have lost had you arrested the progress of science.'

"Whilst we were rising with a progressively increasing speed, we waved our bannerets in token of our cheerfulness, and in order to give confidence to those below who took an interest in our fate. M. Robert made an inventory of our stores; our friends had stocked our commissariat as for a long voyage—champagne and other wines, garments of fur and other articles of clothing.

"'Good,' I said; 'throw that out of the window.' He took a blanket and launched it into the air, through which it floated down slowly, and fell upon the dome of l'Assomption.

"When the barometer had fallen 26 inches, we ceased to ascend. We were up at an elevation of 1,800 feet. This was the height to which I had promised myself to ascend; and, in fact, from this moment to the time when we disappeared from the eyes of our friends, we always kept a horizontal course, the barometer registering 26 inches to 26 inches 8 lines.

"We required to throw over ballast in proportion as the almost insensible escape of the hydrogen gas caused us to descend, in order to remain as nearly as possible at the same elevation. If circumstances had permitted us to measure the amount of ballast we threw over, our course would have been almost absolutely horizontal.

"After remaining for a few moments stationary, our car I changed its course, and we were carried on at the will of the wind. Soon we passed the Seine, between St. Ouen and Asnieres. We traversed the river a second time, leaving Argenteuil upon the left. We passed Sannois, Franconville, Eau-Bonne, St. Leu-Taverny, Villiers, and finally, Nesles. This was about twenty-seven miles from Paris, and we had I reached this distance in two hours, although there was so little wind that the air scarcely stirred.

"During the whole course of this delightful voyage, not the slightest apprehension for our fate or that of our machine entered my head for a moment. The globe did not suffer any alteration beyond the successive changes of dilatation and compression, which enabled us to mount and descend at will. The thermometer was, during more than an hour, between ten and twelve degrees above zero; this being to some extent accounted for by the fact that the interior of the car was warmed by the rays of the sun.

"At the end of fifty-six minutes, we heard the report of the cannon which informed us that we had, at that moment, disappeared from view at Paris. We rejoiced that we had escaped, as we were no longer obliged to observe a horizontal course, and to regulate the balloon for that purpose.

"We gave ourselves up to the contemplation of the views which the immense stretch of country beneath us presented. From that time, though we had no opportunity of conversing with the inhabitants, we saw them running after us from all parts; we heard their cries, their exclamations of solicitude, and knew their alarm and admiration.

"We cried, 'Vive le Roi!' and the people responded. We heard, very distinctly—'My good friends, have you no fear? Are you not sick? How beautiful it is! Heaven preserve you! Adieu, my friends.'

"I was touched to tears by this tender and true interest which our appearance had called forth.

"We continued to wave our flags without cessation, and we perceived that these signals greatly increased the cheerfulness and calmed the solicitude of the people below. Often we descended sufficiently low to hear what they shouted to us. They asked us where we came from, and at what hour we had started.

"We threw over successively frock-coats, muffs, and habits. Sailing on above the Ile d'Adam, after having admired the splendid view, we made signals with our flags, and demanded news of the Prince of Conti. One cried up to us, in a very powerful voice, that he was at Paris, and that he was ill. We regretted missing such an opportunity of paying our respects, for we could have descended into the prince's gardens, if we had wished, but we preferred to pursue our course, and we re-ascended. Finally, we arrived at the plain of Nesles.

"We saw from the distance groups of peasants, who ran on before us across the fields. 'Let us go,' I said, and we descended towards a vast meadow.

"Some shrubs and trees stood round its border. Our car advanced majestically in a long inclined plane. On arriving near the trees, I feared that their branches might damage the car, so I threw over two pounds of ballast, and we rose again. We ran along more than 120 feet, at a distance of one or two feet from the ground, and had the appearance of travelling in a sledge. The peasants ran after us without being able to catch us, like children pursuing a butterfly in the fields.

"Finally, we stopped, and were instantly surrounded. Nothing could equal the simple and tender regard of the country people, their admiration, and their lively emotion.

"I called at once for the cures and the magistrates. They came round me on all sides: there was quite a fete on the spot. I prepared a short report, which the cures and the syndics signed. Then arrived a company of horsemen at a gallop. These were the Duke of Chartres, the Duke of Fitzjames, and M. Farrer. By a very singular chance, we had come down close by the hunting-lodge of the latter.

He leaped from his horse and threw himself into my arms, crying, 'Monsieur Charles, I was first!'

"Charles adds that they were covered with the caresses of the prince, who embraced both of them. He briefly narrated to the Duke of Chartres some incidents of the voyage.

"'But this is not all, monseigneur. I am going away again,' added Charles.

"'What! Going away!' exclaimed the duke.

"'Monseigneur, you will see. When do you wish me to come back again?' I said.

"'In half an hour.'

"'Very well: be it so. In half an hour I shall be with you again.'

"M. Robert descended from the car, and I was alone in the balloon.

"I said to the duke, 'Monseigneur, I go.' I said to the peasants who held down the balloon, 'My friends, go away, all of you, from the car at the moment I give the signal.' I then rose like a bird, and in ten minutes I was more than 3,000 feet above the ground. I no longer perceived terrestrial objects; I only saw the great masses of nature.

"In going away, Charles had taken his precautions against the possible explosion of the balloon, and made himself ready to make certain observations. In order to observe the barometer and the thermometer, placed at different extremities of the car, without endangering the equilibrium, he sat down in the middle, a watch and paper in his left hand, a pen and the cord of the safety-valve in his right.

"I waited for what should happen," continues he. "The balloon, which was quite flabby and soft when I ascended, was now taut, and fully distended. Soon the hydrogen gas began to escape in considerable quantities by the neck of the balloon, and then, from time to time, I pulled open the valve to give it two issues at once; and I continued thus to mount upwards, all the time losing the inflammable air, which, rushing past me from the neck of the balloon, felt like a warm cloud.

"I passed in ten minutes from the temperature of spring to that of winter; the cold was keen and dry, but not insupportable. I examined all my sensations calmly; *I COULD HEAR MYSELF LIVE*, so to speak, and I am certain that at first I experienced nothing disagreeable in this sudden passage from one temperature to another.

"When the barometer ceased to move I noted very exactly eighteen inches ten lines. This observation is perfectly accurate The mercury did not suffer any sensible movement.

"At the end of some minutes the cold caught my fingers; I could hardly hold the pen, but I no longer had need to do so. I was stationary, or rather moved only in a horizontal direction.

"I raised myself in the middle of the car, and abandoned myself to the spectacle before me. At my departure from the meadow the sun had sunk to the people of the valleys; soon he shone for me alone, and came again to pour his rays upon the balloon and the car. I was the only creature in the horizon in sunshine—all the rest of nature was in shade. Ere long, however, the sun disappeared, and thus I had the pleasure of seeing him set twice in the same day. I contemplated for some moments the mists and vapours that rose from the valley and the rivers The clouds seemed to come forth from the earth, and to accumulate the one upon the other. Their colour was a monotonous grey—a natural effect, for there was no light save that of the moon.

"I observed that I had tacked round twice, and I felt currents which called me to my senses. I found with surprise the effect of the wind, and saw the cloth of my flag: extended horizontally.

"In the midst of the inexpressible pleasure of this state of ecstatic contemplation, I was recalled to myself by a most extraordinary pain which I felt in the interior of the ears and in the maxillary glands. This I attributed to the dilation of the air contained in the cellular tissue of the organ as much as to the cold outside. I was in my vest, with my head uncovered. I immediately covered my head with a bonnet of wool which was at my feet, but the pain only disappeared with my descent to the ground.

"It was now seven or eight minutes since I had arrived at this elevation, and I now commenced to descend. I remembered the promise I had made to the Duke of Chartres, to return in half an hour. I quickened my descent by opening the valve from time to time. Soon the balloon, empty now to one half, presented the appearance of a hemisphere.

"Arrived at twenty-three fathoms from the earth, I suddenly threw over two or three pounds of ballast, which arrested my descent, and which I had carefully kept for this purpose. I then slowly descended upon the ground, which I had, so to speak, chosen."

Such is the narrative of the second aerial voyage. After such a memorable ascent one is astonished to learn that Professor Charles never repeated his experiment. It has been said that, in descending from his car, he had vowed that he would never again expose himself to such perils, so strong had been the alarm he felt when the peasants ceasing to hold him down he shot up into the sky with the rapidity of an arrow. But after him a thousand others have followed the daring example he set. With this ascent the memorable year 1783 closed, and the seed which had been sown soon began to be productive.

PART II.

Chapter I. The History of Aerostation from the Year 1783.

> The Open Route—Travels and Travellers—Great Increase in the Number of Air Voyages—Lyons, Ascent of "Le Flesselles— Milan, Ascent of Adriani—Flight of a Balloon from London— Lost Balloons in the Chief Towns of Europe

From the year 1783, in which aerostation had its birth, and in which it was carried to a degree of perfection, beside which the progress of aeronauts in our days seems small, a new route was opened up for travellers. The science of Montgolfier, the practical art of Professor Charles, and the courage of Roziers, subdued the scepticism of those who had not yet given in their adhesion to the possible value of the great discovery, and throughout the whole of France a feverish degree of enthusiasm in the art manifested itself Aerial excursions now became quite fashionable. Let it be understood that we do not here refer to ascents in fixed balloons, that is, in balloons which were attached to the earth by means of ropes more or less long.

M. Biot narrates that, in his young days, when aeronautic ascents were less known than they are in these times, there was in the plain of Grenelle, at the mill of Javelle, an establishment where balloons were constantly maintained for the accommodation of amateurs of both sexes who wished to make ascents in what were called "ballons captifs," or balloons anchored, so to speak, to the earth by means of long ropes They were for a considerable time the rage of fashionable society, and it is not recorded that any accidents resulted from the practice. Of course it may be easily understood with these safe balloons the adventurous aeronauts never ascended to any great height. The reader will find this subject treated under the chapter of military aerostation.

We are at present specially engaged with the narrative of the first attempts in aerostation—the first experiments in the new discovery. We have followed with interest the exciting details of the first adventurous ascents, in which the genius of man first essayed the unexplored paths of the heavens. Yet a continued record of aerial voyages would not be of the same interest. The results of subsequent expeditions, and the impressions of subsequent aeronauts are the same as those already described, or differ from them only in minor points. No important advance is recorded in the art. We shall therefore endeavour not to confine ourselves to the narrative of a dry and monotonous chronology, but to select from the number of ascents that have taken place within the last eighty years, only those whose special character renders them worthy of more detailed and severe investigation.

In order to give an idea of the rapid multiplication of aeronautic experiments, it will suffice to state that the only aeronauts of 1783 are Roziers, the Marquis d'Arlandes, Professor Charles, his collaborateur the younger Robert, and a carpenter, named Wilcox, who made ascents at Philadelphia and London.

A number of balloons were remarkable for the beauty and elegance which we have already spoken of. Among the most beautiful we may mention the "Flesselles" balloon and Bagnolet's balloon.

Of the ascents which immediately succeeded those that have been treated in the first part of our volume, and which are the most memorable in the early annals of aerostation, that of the 17th of January, 1784, is remarkable. It took place at Lyons. Seven persons went into the car on this occasion—Joseph Montgolfier, Roziers, the Comte de Laurencin, the Comte de Dampierre, the Prince Charles de Ligne, the Comte de Laporte d'Anglifort, and Fontaine, who threw himself into the car when it had already begun to move.

A most minute account of this experiment is given in a letter of Mathon de la Cour, director of the Academy of Sciences at Lyons:—"After the experiments of the Champ de Mars and Versailles had become known," he says, "the citizens of this town proposed to repeat them and a subscription was opened for this purpose. On the arrival of the elder Montgolfier, about the end of September, M. de Flesselles, our director, always zealous in promoting whatever might be for the welfare of the province and the advancement of science and art, persuaded him to organise the subscription. The aim of the experiment proposed by Montgolfier was not the ascent of any human being in the balloon. The prospectus only announced that a balloon of a much larger size than any that had been made would ascend—that it would rise to several thousand feet, and that, including the animals that it was proposed it should carry, it would weigh 8,000 lbs. The subscription was fixed at L12, and the number of subscribers was 360."

It was on these conditions that Montgolfier commenced his balloon of 126 feet high and 100 feet in diameter, made of a double envelope of cotton cloth, with a lining of paper between. A strength and consistency was given to the structure by means of ribbons and cords.

The work was nearly finished when Roziers went up in his fire-balloon from La Muette. Immediately the Comte de Laurencin pressed Montgolfier to allow him to go up in the new machine. Montgolfier was only too glad of the opportunity—refused up to this time by the king—of going up himself. From thirty to forty people made application to go with the aeronauts; and on the 26th of December, 1678, Roziers, the Comte de Dampierre, and the Comte de Laporte, arrived in Lyons with the same intention. Prince Charles also arrived; and as his father had taken one hundred subscriptions, his claim to go up could not be refused.

But while the public papers were full of ascents at Avignon, Marseilles, and Paris, it is impossible to describe the vexation of Roziers, when he discovered that Montgolfier's new balloon was not intended to carry passengers, and had not been, from the first, constructed with that view. He suggested a number of alterations, which Montgolfier adopted at once.

On the 7th of January, 1784, all the pieces of which the balloon was composed were carried out to the field called Les Brotteaux, outside the town, from which the ascent was to be made. This event was announced to take place on the 10th and at five o'clock on the morning of that day; but unexpected delays occurred, and in the necessary operations the covering was torn in many places.

On the 15th the balloon was inflated in seventeen minutes, and the gallery was attached in an hour—the fire from which the heated air was obtained requiring to be fed at the rate of 5 lbs. of alderwood per minute; but the preparations had occupied so much time, that it was found, when everything was complete, that the afternoon was too far advanced for the ascent to be made. This machine was destined to suffer from endless misfortunes. It took fire while being inflated, and, several days afterwards, it was damaged by snow and rain. Put nothing discouraged Roziers and his companions. Places had been arranged in the gallery for six persons. After the balloon was at last inflated, Prince Charles and the Comes de Laurencin, Dampierre, and Laporte threw themselves into the gallery. They were all armed, and were determined not to quit their places to whoever might come. Roziers, who wished at the last to enjoy a high ascent, proposed to reduce the number to three, and to draw lots for the purpose. But the gentlemen would not descend. The debate became animated. The four voyagers cried to cut the ropes. The director of the Academy, to whom application was made in this emergency, admiring the resolution and the courage of the four gentlemen, wished to satisfy them in their desire. Accordingly the ropes were cut; but at that moment M. Montgolfier and Roziers threw themselves into the gallery. At the same time a certain M. Fontaine, who had had much to do in the construction of the machine, threw himself in, although it had not previously been arranged that he should be of the party. His boldness in jumping in was pardoned, on the ground of his services and his zeal.

In going away the machine turned to the south-west, and bent a little. A rope which dragged along the ground seemed to retard its ascent; but some intelligent person having cut this with a hatchet, it began to right itself and ascend. At a certain height it turned to the north east. The wind was feeble, and the progress was slow, but the imposing effect was indescribable. The immense machine rose into the air as by some effect of magic. Nearly 100,000 spectators were present, and they were greatly excited at the view. They clapped their hands and stretched their arms towards the sky; women fainted away, or (for some reasons best known to themselves) found relief for their excitement in tears; while the men, uttering cries of joy, waved their handkerchiefs, and threw their hats into the air.

The form of the machine was that of a globe, rising from a reversed and truncated cone, to which the gallery was attached. The upper part was white, the lower part grey; and the cone was composed of strips of stuff of different colours. On the sides of the balloon were two paintings, one of which represented History, the other Fame. The flag bore the arms of the director of the Academy, and above it were inscribed the words "Le Flesselles."

The voyagers observed that they did not consume a fourth of the quantity of combustibles after they had risen into the air, which they consumed when attached to the earth. They were in the gayest humour, and they calculated that the fuel they had would keep them floating till late in the evening. Unfortunately, however, after throwing more wood on the fire, in order to get up to a greater altitude, it was discovered that a rent had been made in the covering, caused by the fire by which the balloon had been damaged two or three days previously. The rent was four feet in length; and as the heated air escaped very rapidly by it, the balloon fell, after having sailed above the earth for barely fifteen minutes.

The descent only occupied two or three minutes, and yet the shock was supportable. It was observed that as soon as the machine had touched the earth all the cloth became unfolded in a few seconds, which seemed to confirm the opinion of Montgolfier, who believed that electricity had much to do in the ascent of balloons. The voyagers were got out of the balloon without accident, and were greeted with the most enthusiastic applause.

On the day of the ascent, the opera of "Iphigenia in Aulis" was given, and the theatre was thronged by a vast assemblage, attracted thither in the hope of seeing the illustrious experimentalists. The curtain had risen when M. and Madame de Flesselles entered their box, accompanied by Montgolfier and Roziers. At sight of them the enthusiasm of the house rose to fever pitch. The other voyagers also entered, and were greeted with the same demonstrations. Cries arose from the pit to begin the opera again, in honour of the visitors. The curtain then fell, and when it again rose, after a few moments, the actor who filled the role of Agamemnon advanced with crowns, which he handed to Madame de Flesselles, who distributed them to the aeronauts. Roziers placed the crown that had been given to him upon Montgolfier's head.

When the actress who played the part of Clytemnestra, sung the passage beginning—

"I love to see these flattering honours paid."

The audience at once applied her song to the circumstances, and re-demanded it, which request the actress complied with, addressing herself to the box in which the distinguished visitors sat. The demonstrations of admiration were continued after the opera was over; and during the whole of the night the gentlemen of the balloon ascent were serenaded.

Two days afterwards, Roziers having appeared at a ball, received further proofs of admiration and honours; and when, on the 22nd of January, he departed for Dijon on his return to Paris, he was accompanied as in a triumph by a numerous cavalcade of the most distinguished young men of the city.

There was, however, at Paris, much discontent with the ascent of "Le Flesselles;" and the Journal de Paris de Paris, which notices so enthusiastically the other ascents of that epoch, speaks slightingly of that at Lyons.

The next great ascent took place at Milan, on the 25th of February, 1784, under the direction of the Chevalier Paul Andriani, who had a balloon constructed by the Brothers Gerli, at his own expense. We read that this balloon was 66 feet in diameter, and that the envelope was composed of cloth, lined in the interior with fine paper.

The balloon was not in all respects constructed like that which rose at Lyons. The grating which supported the fire that kept up the supply of hot air was placed at the mouth of the opening. It was made of copper, was six feet in diameter, and was secured by a number of transverse beams of wood. M. Andriani thought it best to place his fire—contrary to general usage—a little way above the mouth of the opening, and he found out that the activity of the fire was in proportion with that of the air which entered and fed it.

In place of making use of a gallery like that employed by Montgolfier, as much to manage the fire as to carry the traveller and the fuel, he substituted a wide basket, suspended by cords to the edge of the opening of the balloon, at such a distance that fuel could be thrown on with the hand without being inconvenienced by the heat.

Everything being in readiness, the machine was carried to Moncuco, the splendid domain of Andriani, where the first experiments were made; for this gentlemen knew that as the populace are impatient, they are also often un-reasonable, and jump to the hastiest and most inconsiderate conclusion when, in witnessing scientific experiments, any of the arrangements happen to be imperfect, and the results in any respect prove unsuccessful.

Andriani did not deceive himself, for, sure enough, his first attempt did not come up to expectation. The reasons for this failure were the too great quantity of air which the fire drew in, and the unsuitable character of the fuel used.

On the 25th of February, 1784, a second attempt was made. The fire was lighted under the machine, at first with dry birch-wood and afterwards with a bituminous composition, ingeniously concocted by one of the Brothers Gerli. In less than four minutes the balloon was completely inflated, and the men employed to hold it down with ropes perceived that it was on the point of rising. The aeronauts then gave the order to let go. Scarcely was the balloon let off, when it gently rose a short distance, and then flew in a horizontal direction towards a palace in the neighbourhood. In order that the structure should not be destroyed on the walls and the roof of the palace, the voyagers heaped on the fuel, and the spectators, who had gathered together from the surrounding villages, then saw this strange vessel of the air rising with rapidity to a surprising height. Such a phenomenon was so astonishing, that those who beheld it could hardly believe their own eyes; and when the balloon disappeared from view, the delight they had manifested was dashed with fear for the fate of the bold aeronauts. The latter, seeing that the balloon was driving through the air towards a range of rocky hills in the neighbourhood, and perceiving, on the other hand, that their stock of combustibles was nearly exhausted, judged it prudent to descend. They diminished their fire, and came gradually down, warning the multitude below of their intention by means of a speaking-trumpet.

In the course of the descent the balloon alighted upon a large tree, to the great peril of the travellers; but as soon as the fire was increased it again mounted and got clear from the branches while the people below, grasping the cords that were hung out to them, guided the machine to the spot which the voyagers indicated. To descend to terra firma was then a comparatively easy matter, and it was safely accomplished. The fire, which in the case of the French balloons had dried, calcined, and almost consumed the upper part of the balloon, had no evil effect upon that of Andriani, which came down looking as fresh as if it had never been used.

The new idea had now passed the frontiers of France, in which it was originally conceived, and among the other nations, as at first in France, the power of the inflated balloon came to be tested everywhere by the construction of small toy globes.

It was just about five months after the first experiment at Annonay—viz., on the 25th of November, 1783—that the first balloon ascended in London. We are informed, in the History of Aerostation by Tiberius Cavallo, that an Italian, Count Zambeccari, who was staying in the English capital, made a balloon of silk, covered with a varnish of oil. Its diameter was ten feet, and its weight eleven pounds. It was gilded for the double purpose of enhancing its appearance and preventing the escape of air. After having been exposed to public inspection for several days, it was filled three parts full of hydrogen gas, a tin bottle was suspended from it, containing an address to whoever might find it when it should fall, and it was let off from the Artillery Ground, in presence of a vast assembly.

On the 11th of December, 1783, a little balloon, made of goldbeaters' skin, was let off publicly at Turin. This was an experiment similar to that which had been tried at Paris in September. The balloon was seen to penetrate the clouds, then to mount still higher, and finally to disappear entirely in five minutes fifty-four seconds from the time when it was set free.

It was natural, after the experiments made long before with electric paper kites, to employ the balloon in the investigation of the electric conditions of the atmosphere. The first to use it for this purpose was the Abbe Berthelon de Montpellier. He sent up a number of balloons, to which he had attached pieces of metal, long and narrow, and terminating in a cylinder of glass, or other substance suitable for the purpose of isolation, and he obtained sufficient electricity by these means to demonstrate the phenomena of attraction and repulsion, as well as electric sparks.

Cavallo mentions an accident which took place in England about this time, and which served as a warning to all who had to do with balloons filled with hydrogen gas. A balloon thus inflated had been sent up at Hopton, near Matlock, and was found by two men near Cheadle, in Staffordshire. These ingenious persons carried it within doors, and having wished to fully inflate it—half the gas having by this time escaped—they applied a pair of bellows to its mouth. By this means they only forced out the volume of the hydrogen gas that was left; and this gas, coming in contact with a candle that had been placed too near, exploded. The report was louder than that of a cannon, and so powerful was the shock that the men were thrown down, the glass blown out of the windows, and the house otherwise damaged. The men suffered severely, their hair, beards, and eyebrows being completely burnt away, and their faces severely scorched.

At Grenoble, in Dauphine, De Baron let off a balloon on the 13th of January, 1784. It rose, and at first took a northern direction; but, having encountered a current of air, it was carried away in a south-easterly direction, and after flying a distance of three-quarters of a mile, it fell, having traversed this distance in fifteen minutes.

A society, under the presidency of the Abbe de Mably, having constructed a balloon thirty-seven feet high and twenty feet in diameter, sent it off from the court of the Castle of Pisancon, near Romano, on the same day, the 13th of February. At first it was carried to the south by a strong north wind, but after it had risen to 1,000 feet above the surface, its course was changed towards the north. It was calculated that, in less than five minutes, this balloon rose to the height of 6,000 feet.

On the 16th of the same month the Count d'Albon threw off from his gardens at Franconville a balloon inflated with gas, and made of silk, rendered air-tight by a solution of gum-arabic. It was oblong, and measured twenty-five feet in height, and seventeen feet in diameter. To this balloon a cage, containing two guinea-pigs and a rabbit, was suspended. The cords were cut, and the inflated globe rose to an enormous height with the greatest rapidity. Five days afterwards it was found at the distance of eighteen miles, and it is remarkable that, in spite of the cold of the season, and particularly of the elevated region through which the balloon had been passing, the animals were not only living, but in good condition.

On the 3rd of February, 1784, the Marquis de Bullion sent up a paper balloon, of about fifteen feet in diameter. A flat sponge, about a foot square, placed in a tin dish and drenched with a pint of spirits of wine, was the only apparatus made use of to create a supply of heated air. It rose at Paris, and three hours afterwards it was found near Basville, about thirty miles from the capital.

On the 15th of the same month Cellard de Chastelais sent up a paper balloon. Heated air was supplied on this occasion by a paper roll, enclosing a sponge, and soaked in oil, spirits of wine, and grease. A cage, which contained a cat, was attached to this air globe. In thirty-five minutes it had mounted so high that it looked but like the smallest star, and in two hours it had flown a distance of forty-six miles from the place where it was thrown off. The cat was dead, but it was not discovered from what cause.

The first balloon that traversed the English channel was sent off at Sandwich, in Kent, on the 22nd of February, 1784. It was five feet in diameter, and was inflated with hydrogen gas. It rose rapidly, and was carried toward France by a north-west wind. Two hours and a half after it had been let off it was found in a field about nine miles from Lille. The balloon carried a letter, instructing the finder of the balloon to communicate with William Boys, Esq., Sandwich, and to state where and at what time it was found. This request was complied with.

On the 19th of February a similar balloon, five feet in diameter, was sent up from Queen's College, Oxford. It was spherical, and was made of Persian silk, coated with varnish. It was the first balloon sent up from that city.

De Saussure makes mention, in a letter dated from Geneva, the 26th of March, 1784, of certain experiments made in that town with the electricity of the atmosphere by means of fixed balloons—i.e., balloons attached to the earth by ropes, which gave forth sparks and positive electricity.

Mention is also made of a certain M. Argand, of Geneva, who had the honour of making balloon experiments at Windsor in the presence of King George III., Queen Charlotte, and the royal family. About this time (1784) balloons became "the fashion," and frequent instances occur of their being raised by day and night, by means of spirit-lamps, to the great delight of multitudes of spectators.

A letter from Watt to Dr. Lind, of Windsor, dated from Birmingham, 25th December, 1784, narrates an experiment made the summer preceding with a balloon inflated with hydrogen. The balloon was made of fine paper covered with a varnish of oil and filled two-thirds with hydrogen gas, and one-third common air. To the neck of the balloon was attached a sort of squib two feet long, the fuse of which was ignited when the balloon was inflated. The night was calm and dark, and a great multitude was assembled to witness the ascent, which was accomplished with a success that gave delight to all; for, at the end of six minutes the fuse communicated with the squib, and the explosion was like the sound of thunder. The men who saw it from a distance, but were not present at its ascent, took it for a meteor. "Our intention," says Watt, "was, if possible, to discover whether the reverberating sound of thunder was due to echoes or to successive explosions. The sound occasioned by the detonation of the hydrogen gas of the balloon in this experiment, does not enable us to form a definite judgment; all that we can do is to refer to those who were near the balloon, and who affirm that the sound was like that of thunder."

Chapter II. Experiments and Studies — Blanchard at Paris — Guyton de Morveau at Dijon.

The most popular name in aerostation during the Revolution and the Consulate in France is, without doubt, that of Blanchard. We have already referred to him in the chapter which treats of experiments made prior to the discovery of Montgolfier, and we now have to speak of his famous ascent from the Champ de Mars, on the 2nd of March 1784, and of the ascents which followed.

We have seen that he constructed a sort of flying boat, a machine furnished with oars and rigging, with which he managed to sustain himself some moments in the air at the height of eighty feet. This curious machine was exhibited in 1782 in the gardens of the great hotel of the Rue Taranne. But a little time afterwards Montgolfier's discoveries quite altered the conditions under which the aerostatic art was to be pursued. It had no sooner become known than it became public property. The idea was too simple in its grandeur, and was of too easy a kind not to call up a host of imitators. Of these Blanchard was one of the first; but this mechanician was anxious to incorporate his own invention with that of Montgolfier, and he arranged that on the 2nd of March, 1784, he should make an ascent in what he still called his "flying vessel," which he furnished with four wings.

Blanchard and his companion, Pesch, a Benedictine priest, were prevented from going up in the balloon, as represented in our illustration, which was drawn before the event it was intended to commemorate. A certain Dupont de Chambon persisted in accompanying the voyagers. Pushed back by them, he drew his sword, leaped into the car or boat, wounded Blanchard, cut the rigging, and broke the oars or wings. The aeronaut was consequently compelled to have his machine partly re-fitted in great haste, and in the course of a few hours he made the ascent alone in the usual way. Blanchard should have known the uselessness of oars, though he did not abandon their employment in subsequent ascents. The Brothers Montgolfier had dreamed of the employment of oars as a means of guidance, but had ultimately rejected the idea. Joseph wrote to his brother Etienne, about the end of the year 1783:

"For my sake, my good friend, reflect; calculate well before you employ oars. Oars must either be great or small; if great, they will be heavy; if small, it will be necessary to move them with great rapidity. I know no sufficient means of guidance, except in the knowledge of the different currents of air, of which it is necessary to make a study; and these are generally regulated by the elevation." The two brothers often recurred to this idea.

The pictures of the first ascent of Blanchard from the Champ de Mars on the 2nd of March, 1784, in the presence of a vast multitude, show us the oars and the mechanism of his flying-machine fitted to a balloon. The design which we here give seems to us deserving of being considered only as one of the caricatures of the time, especially when we look at the personage dressed in the fool's head-gear, who sits behind and accompanies the triumphant ascent of the aeronaut with music.

It was not with this apparatus that Blanchard effected his ascent, for we have seen that the gearing of his vessel was broken by the infuriated Dupont de Chambon. Yet the aeronaut pretends to have been, to some extent, assisted by his mechanical contrivances. The following is his narrative:—

"I rose to a certain height over Plassy, and perceiving Villette, which I did not despair of reaching in spite of the misfortune that had happened to me, I attached a rope of my rigging to my leg, not being able to make use of my left hand, which I had wrapped in my handkerchief on account of the sword-wound it had received. I fixed up a piece of cloth, and thus made a sort of sail with which I hugged the wind. But the rays of the sun had so heated and rarefied the inflammable air that soon I forgot my rigging in thinking of the terrible danger that threatened me."

Going on to narrate the dangers that beset him, Blanchard describes a number of most extraordinary experiences, which would be better worthy of a place here if they were more like the truth. His curious narrative is thus brought to a close:—

"Escaped from these impetuous and contrary winds, during which I had felt a great degree of cold, I mounted perpendicularly. The cold became excessive. Being hungry I ate a morsel of cake. I wished to drink, but in searching the car nothing was to be seen but the debris of bottles and glasses, which my assailant had left behind him when we were about to depart. Afterwards all was so calm that nothing could be seen or heard. The silence became appalling, and to add to my alarm I began to lose consciousness. I now wished to take snuff, but found I had left my box behind me. I changed my seat many times; I went from prow to stern, but the drowsiness only ceased to assail me when I was struck by two furious winds, which compressed my balloon to such an extent that its size became sensibly diminished to the eye. I was not sorry when I began to descend rapidly upon the river, which at first seemed to me a white thread, afterwards a ribbon, and then a piece of cloth. As I followed the course of the river, the fear that I should have to descend into it, made me agitate the oars very rapidly. I believe that it is to these movements that I owe my being able to cross the river transversely, and get above dry land. When I saw myself upon the plain of Billancourt, I recognised the bridge of Sevres, and the road to Versailles. I was then about as high as the towers above the plain, and I could hear the words and the cries of joy of the people who were following me below. At length I came to a plain about 200 feet in extent. The people then assisted me and brought my vessel to anchor. Immediately I was surrounded by gentlemen and foot passengers who had run together from all parts."

This voyage lasted one hour and a quarter. The most important incident of it was that the balloon was very nearly burst by the expansion of the hydrogen gas. No balloon, as we have already seen, should be entirely inflated at the beginning of a journey. Blanchard had a narrow escape from being the victim of his ignorance of physics, and it is a wonder he was not left to the mercy of fate in a burst balloon, at several thousand feet above the earth.

Biot, the savant, who had watched the experiment, declared that Blanchard did not stir himself, and that the variations of his course are alone to be attributed to the currents of air that he encountered. As he had inscribed upon his flags, his balloons, and his entrance tickets, from which he realised a considerable sum, the ambitious legend, Sic itur ad astra, the following epigram was produced respecting him:—

> From the Field of Mars he took his flight: In a field close by he tumbled; But our money having taken He smiled though sadly shaken, As Sic itur ad astra he mumbled.

What is most important to examine in each of the great aerial voyages that have been made, is the special character which distinguishes them from average experiments. All our great voyages are rendered special and particular by the ideas of the men who undertook them, and the aims which they severally meant to achieve by them. The early ascents of Montgolfier had for their aim the establishment of the fact that any body lighter than the volume of air which it displaces will rise in the atmosphere; those of Roziers were undertaken to prove that man can apply this principle for the purpose of making actual aerial voyages; those of Robertson, Gay-Lussac, &c., were undertaken for the purpose of ascertaining certain meteorological phenomena; those of Conte Coutelle applied aerostation to military uses. A considerable number were made with the view of organising a system of aerial navigation analogous to that of the sea-steerage in a certain direction by means of oars or sails—in a word, to investigate the possibility of sailing through the air to any point fixed upon. It was with this object that the experiments at Dijon took place, and these were the most serious attempts down to our times that have been made to steer balloons.

At the middle of the globe of the balloon were placed four oars, two sails, and a helm and these were under the management of the voyagers, who sat in the car and worked them by means of ropes. The car was also furnished with oars. The report of Guyton de Morveau to the Academy at Dijon informs us that these different paraphernalia were not altogether useless. The following extracts are from this report: —

"The very strong wind which arose immediately before our departure, had driven us down to tee ground many times, making us fear for the safety of our oars, &c., when we resolved to throw over as much ballast as would enable us to rise against the wind. The ballast, including from 70 to 80 lbs. of provisions, was thrown over, and then we rose so rapidly that all the objects around were instantly passed and were very soon lost to view. The swelling form of our balloon told us that the gas inside had expanded under the heat of the sun and the lessening density of the surrounding air. We opened the two valves, but even this outlet was insufficient, and we had to cut a hole about seven or eight inches long in the lower part of the balloon, through which the gas might escape. At five minutes past five we passed above a village which we did not know, and here we let fall a bag filled with bran, and carrying with it a flag and a written message to the effect that we were all well, and that the barometer was recording 20 inches 9 lines, and the thermometer one degree and a half below zero."

Very keen cold attacked the ears, but this was the only inconvenience experienced, until the voyagers were lost in a sea of clouds that shut them out from the view of the earth. The sun at length began to descend, and they then perceived, by a slackening in the lower part of the balloon, that it was time for them to think of returning to the earth. Judging from the compass that they were not far from the town of Auxonne, they resolved to use all their endeavours to reach that place. The sailing appliances had been considerably damaged by the rough weather at starting. The rigging being disarranged, one of the oars had got broken, another had become entangled in the rigging, so that there remained only two of the four oars, and these, being on the same side, were absolutely useless during the greatest part of the voyage. The adventurers, however, assert that they made them work from eight to nine minutes with the greatest ease, making use of them to tack to the south-east.

"We hoped then to be able to descend near where we judged Auxonne to be," the writer continues, "but we lost much gas by the opening in the balloon, and descended more rapidly than we expected or wished. We looked to our small stock of ballast with anxiety, but there was no need of it, and we came very softly down upon a slope."

When the aeronauts arrived at Magny-les-Auxonne, the inhabitants gazed upon them in terror, and two men and three women fell down on their knees before them.

Here is an extract from the report of the experiment of the 12th of June, the principal object of which was the attempt to discover the means of steering in a certain direction:—

"M. de Verley and myself mounted in the balloon," says Guyton de Morveau, "at seven o'clock. We rose rapidly and in an almost perpendicular direction. The fall of the mercury in the barometer was scarcely perceptible when the dilation of the hydrogen gas in the balloon had become considerable. The globe swelled out, and a light vapour around the mouth announced to us that the gas was commencing to escape by the safety-valve. We assisted its escape by pulling the valve-string.

"Having reduced the dilation sufficiently for our purposes, we resolved to attempt the working of the balloon before the whole town and to turn it from the east to the north. We saw with pleasure that our machinery answered By the working of the helm, the prow of our air-boat was turned in the direction we desired. The oars, working only on one side, supported the helm, and altogether we got on as we wished. We described a curve, crossing the road from Dijon to Langres. The mercury had descended to 24 inches 8 lines, which announced that we were gradually rising. We attempted for some time to follow the route to I Langres, but the wind drove us off our course in spite of all our efforts. At nine o'clock our barometer informed us that we had ascended to the height of 6,000 feet. M. de Verley took advantage of this elevation to put some touch wood to a burning-glass 18 lines in diameter, and the touch wood lighted immediately."

The aeronauts decided to direct their course for Dijon. After resetting the helm with this intention, they worked their oars, and proceeded in that direction more than 1,000 feet. But heat and fatigue obliged them to suspend their endeavours, and the current drove them upon Mirebeau, where, throwing out the last of their ballast and regulating their descent, they came softly down upon a corn-field.

The adventurers were cordially welcomed by the ecclesiastics and the magistrates of the place, and after a time they, with their balloon, were carried back on men's shoulders to Dijon.

Chapter III.

> Experiment in Montgolfiers — Roziers and Proust — The Duke of Chartres — The Comte d'Artois — Voyage of the Abbe Carnus to Rodez.

The longest course travelled by Montgolfiere balloons, and the highest elevation reached by them, were achieved by Roziers and Proust with the Montgolfiere la Marie Antoinette, at Versailles, on the 23rd of June, 1784. Roziers himself has left us a picturesque narrative of this excursion from Versailles to Compiegne. He says: —

"The Montgolfiere rose at first very gently in a diagonal line, presenting an imposing spectacle. Like a vessel which has just been precipitated from the stocks, this astonishing machine hung balanced in the air for some time, and seemed to have got beyond human control. These irregular movements intimidated a portion of the spectators, who, fearing that, should there be a fall, their lives would be in danger, scattered away with great speed from under us. After having fed my fire, I saluted the people, who answered me in the most cordial manner. I had time to remark some faces, in which there was a mixed expression of apprehension and joy. In continuing our upward progress, I perceived that an upper current of air made the Montgolfiere bend, but on increasing the heat, we rose above the current. The size of objects on the earth now began perceptibly to diminish, which gave us an idea of the distance at which we were from them. It was then that we became visible to Paris and its suburbs, and so great was our elevation that many in the capital thought we were directly over their heads.

"When we had arrived among the clouds, the earth disappeared from our view. Now a thick mist would envelop us, then a clear space showed us where we were, and again we rose through a mass of snow, portions of which stuck to our gallery. Curious to know how high we could ascend, we resolved to increase our fire and raise the heat to the highest degree, by raising our grating, and holding up our fagots suspended on the ends of our forks.

"Having gained these snowy elevations, and not being able to mount higher, we wandered about for some time in regions which we felt were now visited by man for the first time. Isolated and separated entirely from nature, we perceived beneath us only enormous masses of snow, which, reflecting the sunshine, filled the firmament with a glorious light. We remained eight minutes at this elevation, 11,732 feet above the earth. This situation, however agreeable it might have been to the painter or the poet, promised little to the man of science in the way of acquiring knowledge; and so we determined, eighteen minutes after our departure, to return through the clouds to the earth. We had hardly left this snowy abyss, when the most pleasant scene succeeded the most dreary one. The broad plains appeared before our view in all their magnificence. No snow, no clouds were now to be seen, except around the horizon, where a few clouds seemed to rest on the earth. We passed in a minute from winter to spring. We saw the immeasurable earth covered with towns and villages, which at that distance appeared only so many isolated mansions surrounded with gardens. The rivers which wound about in all directions seemed no more than rills for the adornment of these mansions; the largest forests looked mere clumps or groves, and the meadows and broad fields seemed no more than garden plots. These marvellous tableaux, which no painter could render, reminded us of the fairy metamorphoses; only with this difference, that we were beholding upon a mighty scale what imagination could only picture in little. It is in such a situation that the soul rises to the loftiest height, that the thoughts are exalted and succeed each other with the greatest rapidity. Travelling at this elevation, our fire did not demand continual attention, and we could easily walk about the gallery. We were as much at peace upon our lofty balcony as we should have been upon the terrace of a mansion, enjoying all the pictures which unrolled themselves before us continually, without experiencing any of the giddiness which has disturbed so many persons. Having broken my fork in my exertions to raise the balloon, I went to obtain another one. On my way to get it, I encountered my companion, M. Proust. We ought never to have been on the same side of the balloon, for a capsize and the escape of all our hydrogen gas might have been the result. As it was, so well was the machine ballasted, that the only effect of our being on the one side made the balloon incline a little in that direction. The winds, although very considerable, caused us no uneasiness, and we only knew the swiftness of our progress through the air by the rapidity with which the villages seemed to fly away from under our feet; so that it seemed, from the tranquillity with which we moved, that we were borne along by the diurnal movement of the globe. Often we wished to descend, in order to learn what the people were crying to us the simplicity of our arrangements enabled us to rise, to descend, to move in horizontal or oblique lines, as we pleased and as often as we considered necessary, without altogether landing."

When they came to Luzarche, the delighted aeronauts resolved to land. Already the people were testifying their pleasure at seeing them. Men came running together from all directions, while all the animals rushed away with equal precipitation, no doubt taking the balloon for some wild beast. Finding that their course would lead them straight against certain houses, the aeronauts again increased their fire, and, slightly rising, escaped the buildings that had been in their way. Shortly afterwards they safely landed forty miles from the spot from which they had started.

It was not only the man of science or the mechanician that devoted himself to the task of taking possession of the new empire, but the nobles gave their hands to the aeronauts, and humbly asked the favour of an ascent. The king had addressed letters to the Brothers Montgolfier, and the marvellous invention had become an affair of state. The princes of the blood and the nobles of the court considered it an honour to count among the number of their friends a celebrated aeronaut.

The Count d'Artois, afterwards Charles X., and the Duke de Chartres, father of Louis Philippe, made experiments in aerial navigation. The chemists Alban and Vallet made a magnificent balloon for the Count, who went up many times in it, with several persons of all ranks.

Already at St. Cloud, the Duke of Chartres, afterwards Philippe Egalite, had, on the 15th of July, 1784, made, with the Brothers Robert, an ascent which put their courage to terrible tests. The hydrogen gas balloon was oblong, sixty feet high and forty feet in diameter, and it had been constructed upon a plan supplied by Meunier. In order to obviate the use of the valve, he had placed inside the balloon a smaller globe, filled with ordinary air. This was done on the supposition that, when the balloon rose high, the hydrogen being rarefied would compress the little globe within, and press out of it a quantity of ordinary air equal to the amount of its dilation.

At eight o'clock, the Brothers Robert—Collin and Hullin—and the Duke of Chartres, ascended in presence of an immense multitude. The nearest ranks kneeled down to allow those behind to have a view of the departure of the balloon, which disappeared among the clouds amid the acclamations of the prostrate multitude. The machine, obedient to the stormy and contrary winds which it met, turned several times completely round. The helm, which had been fitted to the machine, and the two oars, gave such a purchase to the winds that the voyagers, already surrounded by the clouds, cut them away. But the oscillations continued, and the little globe inside not being suspended with cords, fell down in such an unfortunate manner as to close up the opening of the large balloon, by means of which provision had been made for the egress of the gas now dilated by the heat of the sun, which poured down its rays, a sudden gust having cleared the space of the clouds. It was feared that the case of the balloon would crack, and the whole thing collapse, in spite of the efforts of the aeronauts to push back the smaller balloon from the opening. Then the Duke of Chartres seized one of the flags they carried, and with the lance-head pierced the balloon in two places. A rent of about nine feet was the consequence, and the balloon began to descend with amazing rapidity. They would have fallen into a lake had they not thrown over 60 lbs. of ballast, which caused them to rise a little, and pass over to the shore, where they got safely to the earth.

The expedition lasted only a few minutes. The Duke of Chartres was rallied by his enemies, who accused him of cowardice; and Monjoie, his historian, making allusion to the combat of Ouessant, says that he had given proofs of his cowardice in the three elements—earth, air, and water.

M. Gray, professor at the seminary of Rodez, presented us some years ago with the following letter from the Abbe Carnus, upon the aerial voyage which he undertook, August 6th, 1784:—

"The progress of the Montgolfiere was so sudden that one might almost have believed that it arose all inflated and furnished out of some chasm in the earth The air was calm, the sky without clouds, the sun very strong. Our fuel and instruments were put into the gallery, my companion, M. Louchet, was at his post, and I took mine. At twenty minutes past eight the cords were loosened, we waved a farewell to the spectators, and while two cannon-shots announced our departure, we were already high above the loftiest buildings.

"To the general acclamations of the crowd succeeded a profound silence. The spectators, half in fear, half in admiration, stood motionless, with eyes fixed, and gazing eagerly at the superb machine, which rose almost vertically with rapidity and also with grandeur. Some women, and even some men, fainted away; others raised their hands to heaven; others shed tears; all grew pale at the sight of our bright fire.

"'We have quitted the earth,' said I to my companion.

"'I compliment you on the fact,' he answered; 'keep up the fire!'

"A truss of hay, steeped in spirits of wine accelerated the swiftness of our ascent. I cast my glance upon the town, which seemed to flee rapidly from under our feet. Terrestrial objects had already lost their shape and size. The burning heat which I felt at first now gave place to a temperature of the most agreeable kind, and the air which we breathed seemed to contain healthful elements unknown to dwellers on the lower earth.

"'How well I am!' I said to Louchet; 'how are you?'

"'As well as can be. Would that I could dispatch a message to the earth!'

"Immediately I threw over a roll of paper on which I had written the words, 'All well on board the City of Rodez.'

"At thirty-two minutes past eight our elevation was at least 6,000 feet above sea level. A flame from our fire, rising from eighteen to twenty feet, sent us up another 1,000 feet. It was then that our machine was seen by every spectator within a circuit of nine miles, and it appeared to be right over the heads of all of them.

"'Send us up out of sight,' said my adventurous confrere.

"I had to moderate his ardour—a larger fire would have burnt our balloon.

"From our moving observatory the most splendid view developed itself. The boundaries of the horizon were vastly extended. The capital of the Rouergue appeared to be no more than a group of stones, one of which seemed to rise to the height of two or three feet. This was no other than the superb tower of the cathedral. Fertile slopes, agreeable valleys, lofty precipices, waste lands, ancient castles perched upon frowning rocks, these form the endlessly varied spectacle which the Rouergue and the neighbouring provinces present to the view of those who traverse the surface of the earth. But how different is the scene to the aerial voyager! We could perceive only a vast country, perfectly round, and seemingly a little elevated in the middle, irregularly marked with verdure, but without inhabitants, without towns, valleys, rivers, or mountains. Living beings no longer existed for us; the forests were changed into what looked like grassy plains; the ranges of the Cantal and the Cevennes had disappeared; we looked in vain for the Mediterranean, and the Pyrenees seemed only a long series of piles of snow, connected at their bases. Our own balloon, which from Rodez appeared about the size of a marble, was the only object that for us retained its natural dimensions. What wonderful sensations then arose within us! I had often reflected upon the works of nature; their magnificence had always filled me with admiration. In this soul-stirring moment how beautiful did nature seem—how grand! With what majesty did it strike my imagination. Never did man appear to me before such an excellent being His latest triumph over the elements recalled to my mind his other conquests of nature. My companion was animated with the same sentiments, and more than once we cried out, 'Vive Montgolfier! Vive Roziers! Vivent ceux qui ont du courage et de la constance!'

"In the meantime our fuel was getting near the end. In eighteen minutes we had run a distance of 12,000 feet. 'Make your observations while I attend to the fire,' said my companion to me. I examined the barometer, the thermometer, and the compass, and having sealed up a small bottle of the air at this elevation, I asked my companion to reduce the fire. We descended 1,800 feet, and at this height I filled another bottle with air.

"Afterwards we felt the refreshing breath of a slight breeze, which carried us gently toward the south-east. In six minutes we had run

18,000 feet. Then, having only sufficient fuel to enable us to choose the place of our descent, we considered whether we should not bring our aerial voyage to a termination. We had neither lake nor forest to fear, and we were secure against danger from fire, as we could detach the grating at some distance from the earth. At fifty-eight minutes past eight all our fuel was exhausted, except two bundles of straw, of four pounds each, which we reserved for our descent. The balloon came gradually down, and terrestrial objects began again to resume their proper forms and dimensions. The animals fled at the sight of our balloon, which seemed likely to crush them in its fall. Horsemen were obliged to dismount and lead their frightened horses. Terrified by such an unwonted sight, the labourers in the fields abandoned their work. We were not more than 600 feet from the earth. We threw on the two bundles of straw, but still gradually descended. The grating was then detached, and I had no difficulty in leaping to the ground. But now a most surprising and unlooked-for event happened. M. Louchet had not been able to descend at the same moment as myself, and the balloon, now free from my weight, immediately re-ascended with the speed of a bird, bearing away my companion. I followed him with my eyes, and it was to my agreeable surprise that I heard him crying to me, 'All is well; fear not!' though it was not without a species of jealousy that I saw him mounting up to the height of 1,400 or 1,500 feet. The balloon, after having run a distance of 3,600 feet in a horizontal direction, began gently to descend at four minutes past nine, at the village of Inieres, after having travelled 42,000 feet from the point of departure. When it had touched the ground it bumped up again two or three feet. M. Louchet jumped out, and seized one of the ropes, but had much difficulty in holding the balloon in hand. He cried to the frightened peasants to come and help him. But they seemed to regard him as a dangerous magician, or as a monster, and they feared to touch the ropes lest they might be swallowed up by the balloon. Soon afterwards I came to the rescue. The balloon was in as thorough repair as when we began our journey. We then pressed out the hot air, folded up the envelope, placed it upon a small cart drawn by two oxen, and drove off with it."

Chapter IV.

> Serio-Comic Aspect of the Subject—The Public Duped—The Abbes Miolan and Janninet at the Luxembourg—Caricatures—The "Minerva" of Robertson, and its Voyage Round the World.

The discovery like that of balloons could not be made public in France without being travestied, and without offering some comic side for the amusement of the wits of the day. Under some old coloured prints, designed with the intention of satirising such unfortunate aeronauts as had collected their money from the spectators, but had failed in inflating their balloons, is written, "The Infallible Means of Raising Balloons"—the infallible means consisting of ropes and pulleys.

While caricature was thus turning its irony upon the efforts of believers in the new idea, serious pamphlets were being written and published with the same object. One of these declares that the discovery is IMMORAL, I. Because since God has not given wings to man, it is impious to try to improve his works, and to encroach upon his rights as a Creator; 2. Because honour and virtue would be in continual danger, if balloons were permitted to descend, at all hours of the night, into gardens and close to windows; 3. Because, if the highway of the air were to remain open to all and sundry, the frontiers of nations would vanish, and property national and personal would be invaded, &c. We do not wish to gather together here the stones which critics threw against the new discovery, unaware all the time that these stones were falling upon their own heads.

It is only fair to state that after the first ascents the public were often duped by pretending aeronauts, whose single aim was to sell their tickets, and who disappeared when the time came for ascending. The result of these frauds was that sometimes honest men were made to suffer as rogues. Even in our own day, when an ascent, seriously intended, fails to succeed, owing to some unforeseen circumstances, the public frequently manifests a decided ill-will to the aeronaut, who is perfectly honest, and only unfortunate.

The famous ascent of the Abbes Miolan and Janninet, at the Luxembourg, may be cited as among the failures which suffered most from the satire of the time. Their immense balloon, constructed at great expense at the observatory, was expected to rise beyond the clouds, and a multitude, each of whom had paid dearly for his ticket, had assembled at the Luxembourg. The morning had been occupied in removing the balloon from the observatory to the place of ascent, and at midday the inflation of it began. The rays of a burning July sun—and one knows what that is in the Luxembourg in Paris—streamed down on the heads of the thousands of spectators. From six in the morning till four in the evening they had waited to see the unheard-of wonder; the ascent, however, was to be so imposing, that nothing could be lost by waiting for it.

But at five in the afternoon the heavy machine was still motionless—inert upon the ground. We need not attempt to describe the scene which took place as the impatience of the multitude increased. Sneers of derision made themselves heard on all sides. A universal murmur, rapidly developing into a clamour, arose amongst the multitude; then, wild with disappointment, the frenzied populace threw themselves upon the barricade, broke it, attacked the gallery of the balloon, the instruments, the apparatus, trampling them under foot, and smashing them in bits. They then rushed upon the balloon and fired it. There was then a general melee. Far from fleeing the fire, every one struggled to seize and carry off a bit of the balloon, to preserve as a relic. The two abbes escaped as they best could, under protection of a number of friends.

After this there fell a perfect shower of lampoons and caricatures. The Abbe Miolan was represented as a cat with a band round its neck, while Janninet appeared as a donkey; and in a coloured print the cat and the ass are shown arriving in triumph upon their famous balloon at the Academy of Montmartre, and are received at the hill of Moulins-a-Vent by a solemn assembly of turkey-cocks and geese in different attitudes. Numerous songs and epigrams, of which the unfortunate abbes were the subjects, also appeared at this time. The letters which composed the words "l'Abbe Miolan" were found to form the anagram, Ballon abime—"the balloon swallowed up."

The most extravagant balloon project was that of Robertson, who published a scheme for making a tour of the world. He called it "La Minerva, an aerial vessel destined for discoveries, and proposed to all the Academies of Europe, by Robertson, physicist" (Vienna, 1804; reprinted at Paris, 1820), Robertson dedicated his project to Volta, and in his dedication he does not scruple to say: "In our age, my friendship seeks only one gratification, that we should both live a sufficiently long time together to enable you to calculate and utilise the results of this great machine, while I take the practical direction of it." The following is this aeronaut's prospectus: —

"There is no limit to the sciences and the arts, which cultivation does not overstep. We have everything to hope and to expect from time, from chance, and from the genius of man. The difference which there is between the canoe of the savage and the man-of-war of 124 guns is perhaps as great as that of balloons as they now are and as they will be in the course of a century. If you ask of an aeronaut why he cannot command the motions of his balloon, he will ask of you in his turn why the inventor of the canoe did not immediately afterwards construct a man-of-war. It must be recollected that there have not yet elapsed forty years since the discovery of the balloon, and that to perfect it would be a work of difficulty, as much from the increased knowledge which such a work would demand, as from the pecuniary sacrifices and the personal devotion which it would involve.

"Thus this invention, after having at first electrified all savants from the one end of the world to the other, has suffered the fate of all discoveries — it was all at once arrested. Did not astronomy wait long for Newton, and chemistry for Lavoisier, to raise them to something like the splendour they now enjoy? Was not the magnet a long time a toy in the hands of the Chinese, without giving birth to the idea of the compass? The electric fluid was known in the time of Thales, but how many ages did we wait for the discovery of galvanism? Yet these sciences, which may be studied in silent retreats, were more likely to yield fruit to the discoverer than aerostatics, which demand courage and skill, and of which the experiments, which are always public, are attended with great cost."

Robertson's proposed machine was to be 150 feet in diameter, and would be capable of carrying 150,000 lbs. Every precaution was to be taken in order to make the great structure perfect. It was to accommodate sixty persons to be chosen by the academics, who should stay in it for several months should rise to all possible elevations, pass through all climates in all seasons, make scientific observations, &c. This balloon, penetrating deserts inaccessible by other means of travel, and visiting places which travellers have never penetrated, would be of immense use in the science of geography: and when under the line, if the heat near the earth should be inconvenient, the aeronauts would, of course, easily rise to elevations where the temperature is equal and agreeable. When their observations, their needs, or their pleasures demanded it, they could descend to within a short distance of the earth, say ninety feet, and fix themselves in their position by means of an anchor. It might, perhaps, be possible, by taking the advantage of favourable winds, to make the tour of the world. "Experience will perhaps demonstrate that aerial navigation presents less inconvenience and less dangers than the navigation of the seas."

The immensity of the seas seemed to be the only source of insurmountable difficulties; "but," says Robertson, "over what a vast space might not one travel in six months with a balloon fully furnished with the necessaries of life, and all the appliances necessary for safety? Besides, if, through the natural imperfection attaching to all the works of man, or either through accident or age, the balloon, borne above the sea, became incapable of sustaining the travellers, it is provided with a boat, which can withstand the waters and guarantee the return of the voyagers."

Such were the ideas promulgated regarding the "Minerva." The following is the serious description given of the machine. The numbers correspond with those on the illustration.

"The cock (3) is the symbol of watchfulness; it is also the highest point of the balloon. An observer, getting up through the interior to the point at which the watchful fowl is placed, will be able to command the best view to be had in the 'Minerva.' The wings at the side (1 and 2) are to be regarded as ornamental. The balloon will be 150 feet in diameter, made expressly at Lyons of unbleached silk, coated within and without with indict-rubber. This globe sustains a ship, which contains or has attached to it all the things necessary for the convenience, the observations, and even the pleasures of the voyagers.

"(a) A small boat, in which the passengers might take refuge in case of necessity, in the event of the larger vessel falling on the sea in a disabled state.

"(b) A large store for keeping the water, wine, and all the provisions of the expedition.

"(cc) Ladders of silk, to enable the passengers to go to all parts of the balloon.

"(e) Closets.

"(h) Pilot's room.

"(1) An observatory, containing the compasses and other scientific instruments for taking the latitude.

"(g) A room fitted up for recreations, walking, and gymnastics.

"(m) The kitchen, far removed from the balloon. It is the only place where a fire shall be permitted.

"(p) Medicine room.

"(v) A theatre, music room, &c.

"—The study.

"(x) The tents of the air-marines, &c. &c."

This balloon is certainly the most marvellous that has ever been imagined—quite a town, with its forts, ramparts, cannon, boulevards, and galleries. One can understand the many squibs and satires which so Utopian a notion provoked.

Chapter V. First Aerial Voyage in England — Blanchard Crosses the Sea in a Balloon.

In spite of their known powers of industry and perseverance, the English did not throw themselves with any great ardour into the exploration of the atmosphere. From one cause or another it is the French and the Italians that have chiefly distinguished themselves in this art. The English historian of aerostation gives some details of the first aerial voyage made in this country by the Italian, Vincent Lunardy.

The balloon was made of silk covered with a varnish of oil, and painted in alternate stripes — blue and red. It was three feet in diameter. Cords fixed upon it hung down and were attached to a hoop at the bottom, from which a gallery was suspended. This balloon had no safety-valve — its neck was the only opening by which the hydrogen gas was introduced, and by which it was allowed to escape.

In September, 1784, it was carried to the Artillery Ground and filled with gas. After being two-thirds filled, the gallery was attached with its two oars or wings, and Lunardy, accompanied by Biggin and Madame Sage, took his place; but it was found that the balloon had not sufficient lifting power to carry up the whole three, and Lunardy went up alone, with the exception of the pigeon, the cat, and the dog, that were with him.

The balloon rose to the height of about twenty feet, then followed a horizontal line, and descended. But the gallery had no sooner touched the earth than Lunardy threw over the sand that served as ballast, and mounted triumphantly, amid the applause of a considerable multitude of spectators. After a time he descended upon a common, where he left the cat nearly dead with cold, ascended, and continued his voyage. He says, in the narrative which he has left, that he descended by means of the one oar which was left to him, the other having fallen over; but, as he states that, in order to rise again, he threw over the remainder of his ballast, it is natural to believe that the descent of the balloon was caused by the loss of gas, because, if he descended by the use of the oar, he must have re-ascended when he stopped using it. He landed in the parish of Standon, where he was assisted by the peasants.

He assures us again that he came down the second time by means of the oar. He says: — "I took my oar to descend, and in from fifteen to twenty minutes I arrived at the earth after much fatigue, my strength being nearly exhausted. My chief desire was to escape a shock on reaching the earth, and fortune favoured me." The fear of a concussion seems to indicate that he descended more because of the weight of the balloon than by the action of the oar.

It appears that the only scientific instrument he had was a thermometer which fell to 29 degrees. The drops of water which had attached themselves to the balloon were frozen.

The second aerial journey in England was undertaken by Blanchard and Sheldon. The latter, a professor of anatomy in the Royal Academy, is the first Englishman who ever went up in a balloon. This ascent was made from Chelsea on the 16th October, 1784.

The same balloon which Blanchard had used in France served him on this occasion, with the difference that the hoop which went round the middle of it, and the parasol above the car, were dispensed with. At the extremity of his car he had fitted a sort of ventilator, which he was able to move about by means of a winch. This ventilator, together with the wings and the helm, were to serve especially the purpose of steering at will, which he had often said was quite practicable as soon as a certain elevation had been reached.

The two aeronauts ascended, having with them a number of scientific and musical instruments, some refreshments, ballast, &c. Twice the ascent failed, and eventually Sheldon got out, and Blanchard went up again alone.

Blanchard says that, on this second ascent, he was carried first north-east, then east-south-east of Sunbury in Middlesex. He rose so high that he had great difficulty in breathing, the pigeon he had with him escaped, but could hardly maintain itself in the rarefied air of such an elevated region, and finding no place to rest, came back and perched on the side of the car. After a time, the cold becoming excessive, Blanchard descended until he could distinguish men on the earth, and hear their shouting. After many vicissitudes he landed upon a plain in Hampshire, about seventy-five miles from the point of departure. It was observed that, so long as he could be clearly seen, he executed none of the feats with his wings, ventilator, &c., which he had promised to exhibit.

Enthusiasm about aerial voyages was now at its climax; the most wonderful deeds were spoken of as commonplace, and the word "impossible" was erased from the language. Emboldened by his success, Blanchard one day announced in the newspapers that he would cross from England to France in a balloon—a marvellous journey, the success of which depended altogether upon the course of the wind, to the mercy of which the bold aeronaut committed himself.

A certain Dr. Jeffries offered to accompany Blanchard. On the 7th of January the sky was calm, in consequence of a strong frost during the preceding night, the wind which was very light, being from the north-north-west. The arranged meets were made above the cliffs of Dover. When the balloon rose, there were only three sacks of sand of 10 lbs. each in it. They had not been long above ground when the barometer sank from 29.7 to 27.3. Dr. Jeffries, in a letter addressed to the president of the Royal Society, describes with enthusiasm the spectacle spread out before him: the broad country lying behind Dover, sown with numerous towns and villages, formed a charming view; while the rocks on the other side, against which the waves dashed, offered a prospect that was rather trying.

They had already passed one-third of the distance across the Channel when the balloon descended for the second time, and they threw over the last of their ballast; and that not sufficing, they threw over some books, and found themselves rising again. After having got more than half way, they found to their dismay, from the rising of the barometer, that they were again descending, and the remainder of their books were thrown over. At twenty-five minutes past two o'clock they had passed three-quarters of their journey, and they perceived ahead the inviting coasts of France. But, in consequence either of the loss or the condensation of the inflammable gas, they found themselves once more descending. They then threw over their provisions, the wings of the car, and other objects. "We were obliged," says Jeffries, "to throw out the only bottle we had, which fell on the water with a loud sound, and sent up spray like smoke."

They were now near the water themselves, and certain death seemed to stare them in the face. It is said that at this critical moment Jeffries offered to throw himself into the sea, in order to save the life of his companion.

"We are lost, both of us," said he; "and if you believe that it will save you to be lightened of my weight, I am willing to sacrifice my life."

This story has certainly the appearance of romance, and belief in it is not positively demanded.

One desperate resource only remained—they could detach the car and hang on themselves to the ropes of the balloon. They were preparing to carry out this idea, when they imagined they felt themselves beginning to ascend again. It was indeed so. The balloon mounted once more; they were only four miles from the coast of France, and their progress through the air was rapid. All fear was now banished. Their exciting situation, and the idea that they were the first who had ever traversed the Channel in such a manner, rendered them careless about the want of certain articles of dress which they had discarded. At three o'clock they passed over the shore half-way between Cape Blanc and Calais. Then the balloon, rising rapidly, described a great arc, and they found themselves at a greater elevation than at any part of their course. The wind increased in strength, and changed a little in its direction. Having descended to the tops of the trees of the forest of Guines, Dr. Jeffries seized a branch, and by this means arrested their advance. The valve was then opened, the gas rushed out, and the aeronauts safely reached the ground after the successful accomplishment of this daring and memorable enterprise.

A number of horsemen, who had watched the recent course of the balloon, now rode up, and gave the adventurers the most cordial reception. On the following day a splendid fete was celebrated in their honour at Calais. Blanchard was presented with the freedom of the city in a box of gold, and the municipal body purchased the balloon, with the intention of placing it in one of the churches as a memorial of this experiment, it being also resolved to erect a marble monument on the spot where the famous aeronauts landed.

Some days afterwards Blanchard was summoned before the king, who conferred upon him an annual pension of 1,200 livres. The queen, who was at play at the gambling table, placed a sum for him upon a card, and presented him with the purse which she won.

Chapter VI. Zambeccari's Perilous Trip Across the Adriatic Sea.

There is not in the whole annals of aerostation a more moving catastrophe than that of the unfortunate Comte Zambeccari, who, during an aerial journey on October the 7th, 1804, was cast away on the waves of the Adriatic.

The history of Zambeccari is dramatic throughout. After having been taken by the Turks and thrown into the Bay of Constantinople, from which he with difficulty escaped, he devoted himself to the study and practice of aerial navigation. He fancied he could make use of a lamp supplied with spirits of wine, the flame of which he could direct at will, in the hope of thus being able to steer the balloon in whatever direction he chose. One day his balloon damaged itself against a tree at Boulogne, and the spirits of wine set his clothes on fire. The flames with which the aeronaut was covered only served to increase the ascending power of the balloon, and the frightened spectators, among whom were Zambeccari's young wife and children, saw him carried up into the clouds out of sight. He succeeded, however, in extinguishing the fire which surrounded him.

In 1804, he organised a series of experiments at Milan, for which he received, in advance, the sum of 8,000 crowns; but the experiments failed, in consequence of the inclemency of the weather, the treachery of his assistants, and the malice of his rivals.

At length, on the 7th of October, after a fall of rain which lasted forty-eight hours, and which had delayed the announced ascent, he resolved, whatever might happen, to carry it out, though all the chances were against him. Eight young men whom he had instructed, and who had promised him their assistance in filling the balloon, failed him at the critical moment. Still, however, he continued his labours, with the help of two companions, Andreoli and Grassetti. Wearied with his long-continued efforts, dis-appointed and hungry, he took his place in the car.

The two companions whom we have named went with him. They rose gently at first, and hovered over the town of Bologna. Zambeccari says, "The lamp, which was intended to increase our ascending force, became useless. We could not observe the state of the barometer by the feeble light of a lantern. The insupportable cold that prevailed in the high region to which we had ascended, the weariness and hunger arising from my having neglected to take nourishment for twenty-four hours, the vexation that embittered my spirit—all these combined produced in me a total prostration, and I fell upon the floor of the gallery in a profound sleep that was like death. 'The same misfortune overtook my companion Grassetti. Andreoli was the only one who remained awake and able for duty—no doubt because he had taken plenty of food and a large quantity of rum. Still he suffered from the cold, which was excessive, and his endeavours to wake me were for a long time vain. Finally, however, he succeeded in getting me to my feet, but my ideas were confused, and I demanded of him, like one newly awaking from a dream, 'What is the news? Where are we? What time is it? How is the wind?'

"It was two o'clock. The compass had been broken, and was useless; the wax light in the lantern would not burn in such a rarefied atmosphere. We descended gently across a thick layer of whitish clouds, and when we had got below them, Andreoli heard a sound, muffled and almost inaudible, which he immediately recognised as the breaking of waves in the distance. Instantly he announced to me this new and fearful danger. I listened, and had not long to wait before I was convinced that he was speaking the truth. It was necessary to have light to examine the state of the barometer, and thus ascertain what was our elevation above the sea level, and to take our measures in consequence. Andreoli broke five phosphoric matches, without getting a spark of fire. Nevertheless, we succeeded, after very great difficulty, by the help of the flint and steel, in lighting the lantern. It was now three o'clock in the morning—we had started at midnight. The sound of the waves, tossing with wild uproar, became louder and louder, and I suddenly saw the surface of the sea violently agitated just below us. I immediately seized a large sack of sand, but had not time to throw it over before we were all in the water, gallery and all. In the first moment of fright, we threw into the sea everything that would lighten the balloon—our ballast, all our instruments, a portion of our clothing, our money, and the oars. As, in spite of all this, the balloon did not rise, we threw over our lamp also. After having torn and cut away everything that did not appear to us to be of indispensable necessity, the balloon, thus very much lightened, rose all at once, but with such rapidity and to such a prodigious elevation, that we had difficulty in hearing each other, even when shouting at the top of our voices. I was ill, and vomited severely. Grassetti was bleeding at the nose; we were both breathing short and hard, and felt oppression on the chest. As we were thrown upon our backs at the moment when the balloon took such a sudden start out of the water and bore us with such swiftness to those high regions, the cold seized us suddenly, and we found ourselves covered all at once with a coating of ice. I could not account for the reason why the moon, which was in its last quarter, appeared on a parallel line with us, and looked red as blood.

"After having traversed these regions for half an hour, at an immeasurable elevation, the balloon slowly began to descend, and at last we fell again into the sea, at about four in the morning I cannot determine at what distance we were from land when we fell the second time. The night was very dark, the sea rolling heavily, and we were in no condition to make observations. But it must have been in the middle of the Adriatic that we fell. Although we descended gently, the gallery was sunk, and we were often entirely covered with water. The balloon being now more than half empty, in consequence of the vicissitudes through, which we had passed, gave a purchase to the wind, which pressed against it as against a sail, so that by means of it we were dragged and beaten about at the mercy of the storm and the waves. At daybreak we looked out and found ourselves opposite Pesaro, four miles from the shore. We were comforting ourselves with the prospect of a safe landing, when a wind from the land drove us with violence away over the open sea. It was now full day, but all we could see were the sea, the sky, and the death that threatened us. Certainly some boats happened to come within sight; but no sooner did they see the balloon floating and striping upon the water than they made all sail to get away from it. No hope was then left to us but the very small one of making the coasts of Dalmatia, which were opposite, but at a great distance from us. Without the slightest doubt we should have been drowned if heaven had not mercifully directed towards us a navigator who, better informed than those we had seen before, recognised our machine to be a balloon and quickly sent his long-boat to our rescue. The sailors threw us a stout cable, which we attached to the gallery, and by means of which they rescued us when fainting with exposure. The balloon thus lightened, immediately rose into the air, in spite of all the efforts of the sailors who wished to capture it. The long boat received a severe shock from its escape, as the rope was still attached to it, and the sailors hastened to cut themselves free. At once the balloon mounted with incredible rapidity, and was lost in the clouds, where it disappeared for ever from our view. It was eight in the morning when we got on board. Grassetti was so ill that he hardly showed any signs of life. His hands were sadly mutilated. Cold, hunger, and the dreadful anxiety had completely prostrated me. The brave captain of the vessel did everything in his power to restore us. He conducted us safely to Ferrara, whence we were carried to Pola, where we were received with the greatest kindness, and where I was compelled to have my fingers amputated."

Chapter VII. Garnerin — Parachutes — Aerostation at Public Fetes.

"On the 22nd October, 1797," says the astronomer Lalande, "at twenty-eight minutes past five, Citizen Garnerin rose in a balloon from the park of Monceau. Silence reigned in the assembly, anxiety and fear being painted on the visages of all. When he had ascended upwards of 2,000 feet, he cut the cord that connected his parachute and car with the balloon. The latter exploded, and Garnerin descended in his parachute very rapidly. He made a dreadful lurch in the air, that forced a sudden cry of fear from the whole multitude, and made a number of women faint. Meanwhile Citizen Garnerin descended into the plain of Monceau; he mounted his horse upon the spot, and rode back to the park, attended by an immense multitude, who gave vent to their admiration for the skill and talent of the young aeronaut. Garnerin was the first to undertake this most daring and dangerous venture. He had conceived the idea of this feat while lying a prisoner of state in Buda, Hungary." Lalande adds that he went and announced his success at the Institute National, which was assembled at the time, and which listened to him with the greatest interest.

Robertson conducted an experiment of descending by means of a parachute at Vienna, in 1804, in which he received all the glory, without partaking of any of the danger. He made the public preparations for an ascent in the balloon, his pupil, Michaud, however, took his place in the car, and made the ascent.

Robertson says that on this occasion he yielded to the entreaties of a young man who was his pupil, and had begged to be allowed to make his debut before such a great multitude. In this case a slight improvement was made in the parachute. The car was surrounded by a cloth of silk, which, when the aeronaut cut himself away from the balloon, spread itself out in such a way as to form a second parachute.

Robertson made all the preparations, and Michaud had no more to do than place himself in the car. Loud applause arose on all sides. Michaud had ascended 900 feet above the earth when the signal for his cutting himself clear of the balloon was given, by the firing of a cannon. He at once cut the two strings, and the balloon soared away into the upper regions, whilst he was left for one terrible moment to fate. The fall was at first rapid, but the two parachutes soon opened themselves simultaneously, and presented a majestic appearance. In a few seconds the aeronaut had traversed the space that intervened between him and the assembly, and found himself safely landed on the ground, at a short distance from the place whence he had set out, while the whole air was rent with shouts of applause. This experiment was deemed a most extraordinary one. Compliments were showered upon Robertson from all sides, and the court presented him with rich presents.

Balloons have always formed a prominent feature at the fetes of Paris, for the celebration of the chief events of the Revolution, the Consulate, and the Empire — the first of these epochs being that in which these aerial vessels were held in highest esteem.

Jacques Garnerin had played a brilliant role as aeronaut under the Directory, the Consulate, and the Empire; and it was he who after the coronation of the Emperor Napoleon I., was charged with the raising of a monster balloon, which was arranged to ascend, with the accompaniment of fireworks, on the evening of the 16th of December, 1804.

An uncommon incident connected with this event serves to show us the spirit of fatalism with which the character of Napoleon I. was infected. "The Man of Destiny" believed in the destiny of man; he had faith in his star alone; and from the height of his greatness the new ruler, consecrated emperor and king by the Pope, beheld a presage of misfortune in a chance circumstance, insignificant to all but himself, in the experiment of which we are about to recount the history.

The fete given by the city of Paris to their majesties embraced the whole town, from the Champs Elysees to the Barriere du Trone, on the square of the Hotel de Ville. Upon the river throughout its length between the Isle of St. Louis and the bridge of Notre Dame, an immense display of fireworks was to take place. The scene to be represented was the passage of Mont St. Bernard. Garnerin was stationed with his balloon in front of the gate of the church of Notre

Dame. At eleven o'clock in the evening, at the moment when the first discharge of fireworks made the air luminous with a hundred thousand stars, Garnerin threw off his immense balloon. The chief feature of it was the device of a crown, designed in coloured lanterns arranged round the globe. It rose splendidly, and with the most perfect success.

On the following morning the inhabitants of Rome were astounded to behold advancing toward them from the horizon a luminous globe, which threatened to descend upon their city. The excitement was intense. The balloon passed the cupola of St. Peter's and the Vatican; then descending, it touched the ground, but rose again, and finally it sank into the wafers of Lake Bracciano.

It was drawn from the water, and the following inscription, emblazoned in letters of gold upon its vast circumference, was printed, published, and read throughout the whole of Italy—"Paris, 25eme Primaire, an XIII., couronnement de l'empereur Napoleon, 1er par S.S. Pie VII."

In touching the earth, the balloon happened to strike against the tomb of the Emperor Nero, and, owing to the concussion, a portion of the crown was left upon this ancient monument. The Italian journals, which were not so strictly under the supervision of the government as were the journals of France, gave the full particulars of these minor events; and certain of them, connecting the names of Nero and Napoleon, indulged in malicious remarks at the expense of the French emperor. These facts came to the ear of the great general, who manifested much indignation, dismissed the innocent Garnerin from his post, and appointed Madame Blanchard to the supervision of all the balloon ascents which took place at the public fetes.

The balloon was preserved in the vaults of the Vatican in Rome, accompanied with an inscription narrating its travels and wonderful descent—minus the circumstance of the tomb. It was removed, as might be supposed, in 1814. From this time the ascents of balloons took place for the most part only on the occasions of coronations and other great public fetes.

Chapter VIII. Green's Great Journey Across Europe.

It is probable that at the origin of navigation, man, before he had invented oars and sails, made use of trunks of trees upon which he trusted himself, leaving the rest to the winds and the currents of the water, whether these were known or unknown. There is some analogy between such rude rafts, the first discovered means of navigation on water, and balloons, the first discovered means of navigation in air. But unquestionably the advantage is with the latter. No means have yet been found of directly steering balloons, but by allowing the gas to escape the aeronaut can descend at will, and by lightening his car of part of the ballast he carries he can ascend as readily. It must also be remembered that the currents of air vary in their directions, according to their elevation, and were the aeronaut perfectly acquainted with aerial currents, he might, by raising or lowering himself, find a wind blowing in the direction in which he wished to proceed, and the last problem of aerostation would be solved. That any such knowledge can ever be acquired it is impossible to say; but this much may with safety be advanced, that distant journeys may frequently be taken with balloons for useful purposes.

One of the most remarkable excursions of this kind was that superintended by Green, in 1836, from London to Germany. This journey, 1,200 miles in length, is the longest that has been yet accomplished. Green set out from London on the 7th of November, 1836, accompanied by two friends—Monk-Mason, the historian of the journey, and a gentleman named Molland. Not knowing to what quarter of the globe he might be blown, Green provided himself with passports to all the states of Europe, and with a quantity of provisions sufficient to last him for some time, should he be driven by the wind over the sea. Shortly after mid-day the balloon rose with great grandeur, and, urged by a light breeze, floated to the south-east, over the plains of Kent. At four o'clock the voyagers sighted the sea.

"It was forty-eight minutes past four," says Monk-Mason, "that we first saw the line of waves breaking on the shores beneath us. It would have been impossible to have remained unmoved by the grandeur of the spectacle that spread out before us. Behind us were the coasts of England, with their white cliffs half lost in the coming darkness. Beneath us on both sides the ocean spread out far end wide to where the darkness closed in the scene. Opposite us a barrier of thick clouds like a wall, surmounted all along its line with projections like so many towers, bastions, and battlements, rose up from the sea as if to stop our advance. A few minutes afterwards we were in the midst of this cloudy barrier, surrounded with darkness, which the vapours of the night increased. We heard no sound. The noise of the waves breaking on the shores of England had ceased, and our position had for some time cut us off from all the sounds of earth."

In an hour the Straits of Dover were cleared, the lights of Calais shone out toward the voyagers, and the sound of the town drums rose up toward them. "Darkness was now complete," continues the writer, "and it was only by the lights, sometimes isolated, sometimes seen in masses, and showing themselves far down on the earth beneath us, that we could form a guess of the countries we traversed, or of the towns and villages which appeared before us every moment. The whole surface of the earth for many leagues round showed nothing but scattered lights, and the face of the earth seemed to rival the vault of heaven with starry fires. Every moment in the earlier part of the night before men had betaken themselves to repose, clusters of lights appeared indicating large centres of population.

"Those on the horizon gave us the notion of a distant conflagration. In proportion as we approached them, these masses of lights appeared to increase, and to cover a greater space, until, when right over them, they seemed to divide themselves into different parts, to stretch out in long streets, and to shine in starry quadrangles round the squares, so that we could see the exact plan of each city, given as on a small map. It would be difficult to give an idea of what sort of effect such a scene in such circumstances produces. To find oneself transported in the darkness of night, in the midst of vast solitudes of air, unknown, unperceived, in secret and in silence, exploring territories, traversing kingdoms, watching towns which come into view, and pass out of it before one can examine them in detail— these circumstances are enough in themselves to render sublime a science which, independent of these adjuncts, would be so interesting. If you add to this the uncertainty which, increasing as we went on into the night, began to assail us respecting our voyage, our ignorance of where we were, and what were the objects we were attempting to discover, you may form some idea of our singular position."

About midnight, the travellers found themselves above Liege. Situated in the midst of a thickly-peopled country, full of foundries, smelting works, and forges, this town was quite a blaze of light. The gas-lamps with which this town is so well lighted, clearly marked out for our travellers the main streets, the squares, and the public buildings. But after midnight, at which time the lamps in continental towns are mostly put out, the whole of the under world disappeared from the view of the aeronauts.

"After the turn of the night," says Mason, "the moon did not show itself, and the heavens, always more sombre when regarded from great altitudes, seemed to us to intensify the natural darkness. On the other hand, by a singular contrast, the stars shone out with unusual brilliancy, and seemed like living sparks sown upon the ebony vault that surrounded us. In fact, nothing could exceed the intensity of the night which prevailed during this part of our voyage. A black profound abyss surrounded us on all sides, and, as we attempted to penetrate into the mysterious deeps, it was with difficulty we could beat back the idea and the apprehension that we were making a passage through an immense mass of black marble, in which we were enclosed, and which, solid to within a few inches of us, appeared to open up at our approach."

Until three o'clock the voyagers were in this state. The height of the balloon, as calculated by the barometer, was 2,000 feet. They had not then anything to fear from a disastrous encounter, when all at once a sudden explosion was heard, the silk of the balloon quivered, the car received a violent shock, and seemed to be shot suddenly into the gloomy abyss. A second explosion and a third succeeded, accompanied each time by this fearful shock to the car. The travellers soon found out that, owing to the great altitude, the gas had expanded, and the rope which surrounded it, saturated with water, and frozen with the intense cold, had yielded to the pressure, in jerks which caused the report and the shock.

"From time to time," continues Mason, "vast masses of clouds covered the lower regions of the atmosphere, and spread a thick, whitish veil over the earth, intercepting our view, and leaving us for some time uncertain if this was not a continuation of the same plains covered with snow which we had already noticed. From these masses of vapour, there seemed more than once during the night to come a sound as of a great fall of water, or the contending waves of the sea; and it required all the force of our reason, joined to our knowledge—such as it was—of the direction of our route, to repress the idea that we were approaching the sea, and that, driven by the wind, we had, been carried along the coasts of the North Sea or the Baltic. As the day advanced these apprehensions disappeared. In place of the unbroken surface of the sea, we gradually made out the varied features of a cultivated country, in the midst of which flowed a majestic river, which lost itself, at both extremities, in the mist that still lay on the horizon."

This river was the Rhine, and as the neighbourhood seemed suitable for a descent, and as the travellers did not wish to be carried too far into the heart of Europe, they allowed a portion of the gas to escape, came gradually down, and dropped their anchor.

It was then half-past seven in the morning. It was only then that the inhabitants, who had hitherto held themselves aloof, watching the movements of the strangers from under the brushwood, began to assemble from all sides. A few words in German spoken from the balloon dissipated their fears, and, recovering from their mistrust, they hastened immediately to lend assistance to the aeronauts The latter were now informed that the place they had selected for their descent was in the Duchy of Nassau. The town of Wiberg, where Blanchard had descended, after his ascent at Frankfort in 1785 was, by a singular chance, only two leagues distant. The three aeronauts received a most flattering reception, and, in memory of the event, they placed the flag which they had borne in their car during their adventurous excursion in the ducal palace, side by side with that of Blanchard.

"Thus," says Mason, "terminated an expedition which, whether we regard the extent of the journey, the length of time occupied in it, or the results which were the objects of the experiment, may justly be considered as one of the most interesting and most important ever undertaken. The best answer which one could give to those who would be disposed to criticise the employment of the peculiar means which we made use of, or to doubt their efficiency, would be to state that, after having traversed without hindrance, without either danger or difficulty, so large a portion of the European continent, we arrived at our destination still in possession of as much force as, had we wished it, might have carried us round the whole world."

Chapter IX. The "Geant" Balloon.

Not a few of our readers will remember the ascent of Nadar's colossal balloon from Paris, on Sunday, the 18th of October, 1863. This balloon was remarkable as having attached to it a regular two-story house for a car. Its ascent was witnessed by nearly half a million of persons. The balloon, after passing over the eastern part of France, Belgium, and Holland, suffered a disastrous descent in Hanover the day after it started on its perilous journey. It was a fool-hardy enterprise to construct such a gigantic and unmanageable balloon, presenting such an immense surface to the atmosphere, and being so susceptible to adverse aerial currents as to become the helpless prey of the elements; and it was still more fool-hardy to place the lives of its passengers at the mercy of such terrible and ungovernable forces. A large section of the public laboured under the delusion that Nadar's balloon was one capable of being steered. In reality, however, the 'Geant' was unquestionably the most rebellious and unruly specimen of its class that has been made since the days of Montgolfier. The object in view when this formidable monster was designed and constructed was to create the means to collect sufficient funds to form a "Free Association for Aerial Navigation by means of MACHINES HEAVIER THAN AIR," and for the construction of machines on this principle. The receipts from the exhibition of the "Geant" were intended to form the first capital of the association. The hopes, however, of the promoters have not been realised in this respect; for while the expenses of the construction of the balloon have amounted, directly and indirectly, to the sum of L8,300, its two ascents in Paris and its exhibition in London produced only L3,300.

Space forbids us to enter at length on the various stages of the idea of aerial navigation by means of an apparatus heavier than the atmosphere. The idea is not, however, by any means so absurd as it appears at first sight. Those who, like Arago, declare that the word "impossible" does not exist, except in the higher mathematics, and those who look hopefully to the future instead of resting content with the past, will join in applauding the spirit which dictated the manifesto of aerial locomotion to the founder of the association which we are about to describe. M. Babinet, speaking on this subject before the French Polytechnic Association, said: "It is absurd to talk of guiding balloons. How will you set about it? How is it possible that a balloon—say, for instance, like the Flesselles, whose diameter measures 120 feet—can resist and manoeuvre against opposing winds or currents of air? It would require a power equal to 400 horses for the sails of a ship to struggle on equal terms with the wind. Suppose an impossibility, namely, that a balloon could carry with it a force equal to 400 horse-power; this result would be of little use, for under the immense weight the fragile covering of the balloon would instantly collapse. If all the horses of a regiment were harnessed to the car of a balloon by means of a long rope, the result would be that the balloon would fly into shivers, being too fragile to withstand these two opposing forces. Man must seek to raise himself in the air by another mode of operation altogether, if he wish to guide himself at the same time. Some time ago I bought a play thing, very much in vogue at that time, called a Stropheor. This toy was composed of a small rotating screw propeller, which revolved on its own support when the piece of string wound round it was pulled sharply. The screw was rather heavy, weighing nearly a quarter of a pound, and the wings were of tin, very broad and thick. This machine, however, was rather too eccentric for parlour use, for its flight was so violent that it was continually breaking the pier glass, if there was one in the room; and, failing this, it next attacked the windows. The ascending force of this machine is so great that I have seen one of them fly over Antwerp Cathedral, which is one of the highest edifices in the world. The air from underneath the machine is exhausted by the action of the screw, which, passing under the wings, causes a vacuum, while the air above it replenishes and fills this void, and under the influence of these two causes the apparatus mounts from the earth. But the problem is not solved by means of this plaything, whose motive power is exterior to it. Messrs. Nadar, Ponton, D'Amecourt, and De la Landelle teach us better than this, although the wings of their different models are entirely unworthy of men who desire to demonstrate a truth to short-lived mortals. We have only arrived as yet at the infancy of the process, but we have made a good beginning, for, having once proved that a machine capable of raising itself in the air, wholly unaided from without, can be made, we have overcome with this apparently small result the whole difficulty. The principle of propulsion by means of a screw is by no means a novelty. It was first utilised in windmills, whose sails are nothing more nor less than an immense screw which is turned by the action of the wind on its surface. In the case of turbine water-wheels, where perhaps 970 cubic feet of water are utilised by means of a mechanism not larger than a hat, we see another illustration of it, with this difference, that water takes the place of wind as the motive power.

"The aerial screw is beset with great difficulties, but if we can succeed through its agency in raising even the smallest weight, we may be confident of being able to raise a heavier one, for a large machine is always more powerful in proportion to its size than a small one.

"Mlle. Garnerin once made a bet that she would guide herself in her descent from a considerable altitude towards a fixed spot on the earth at some distance, with no other help than the parachute; and she was really able to guide herself to within a few feet of the specified spot, by simply altering the inclination of the parachute.

"From observations in mountainous districts, where large birds of prey may be seen to the best advantage hovering with outstretched wings, I have come to the conclusion that they first of all attain the requisite height and then, extending their wings in the form of a parachute, let themselves glide gradually towards the desired spot. Marshal Niel confirms this opinion by his experience in the mountains of Algeria. It is, therefore, clear from these examples that we should possess the power of transporting ourselves from place to place if we could only discover a means of raising a weight perpendicularly in the air, which would then act as a capital of power, only requiring to be expended at will."

From the foregoing remarks we may gather an idea of the importance which may be attached to aerial locomotion notwithstanding the successive failures of all those who have hitherto taken up the subject. We come now to the description of the memorable ascent of the 'Geant.'

We learn from the very interesting account of the 'Geant,' published at the time, all the mishaps and adventures it outlived from the time of the first stitch in its covering to its final inflation with gas. We must, however, be content to take up the narrative at the point at which the 'Geant,' with thirteen passengers on board, had, in obedience to the order to "let go," been released from the bonds which held it to the earth. The narrative is, as our readers will perceive, written in somewhat exaggerated language:—

"The 'Geant' gave an almost imperceptible shake on finding itself free, and then commenced to rise. The ascent was slow and gradual at first—the monster seemed to be feeling its way. An immense shout rose with it from the assembled multitude. We ascended grandly, whilst the deafening clamour of two hundred thousand voices seemed to increase. We leant over the edge of the car, and gazed at the thousands of faces which were turned towards us from every point of the vast plain, in every conceivable angle of which we were the common apex. We still ascended. The summits of the double row of trees which surround the Champ de Mars were already under us. We reached the level of the cupola of the Military School. The tremendous uproar still reached us. We glided over Paris in an easterly direction, at the height of about six hundred feet. Every one took up the best possible position on the six light cane stools, and on the two long bunks at either end of the car, and contemplated the marvellous panorama spread out under us, of which we never grew weary.

"There is never any dizziness in a balloon, as is often erroneously supposed, for in it you are the only point in space without any possibility of comparison with another, and therefore the means of becoming giddy are not at hand."

A very experienced aeronaut, who numbers his ascents by hundreds, has assured me that he never knew of a single case of dizziness.

"The earth seems to unfold itself to our view like an immense and variegated map, the predominant colour of which is green in all its shades and tints. The irregular division of the country into fields made it resemble a patchwork counterpane. The size of the houses, churches, fortresses, was so considerably diminished as to make them resemble nothing so much as those playthings manufactured at Carlsruhe. This was the effect produced by a microscopic train, which whistled very faintly to attract our attention, and which seemed to creep along at a snail's pace, though doubtless going at the rate of thirty miles an hour, and was enveloped in a minute cloud of smoke. What a lasting impression this microscopic neatness makes on us! What is that white puff I see down there? the smoke of a cigar? No: it is a cloud of mist. It must be a perfect plain that we are looking at, for we cannot distinguish between the different altitudes of a bramble-bush and an oak a hundred years old!

"It is one of the delights of an aeronaut to gaze on the familiar scenes of earth from the immense height of the car of a balloon! What earthly pleasure can compare with this! Free, calm, silent, roving through this immense and hospitable space, where no human form can harm me, I despise every evil power; I can feel the pleasure of existence for the first time, for I am in full possession, as on no other occasion, of perfect health of mind and body. The aeronauts of the 'Geant' will scarcely condescend to pity those miserable mortals whom they can only faintly recognise by their gigantic works, which appear to them not more dignified than ant-hills!

"The sun had already set behind the purple horizon in our rear. The atmosphere was still quite clear round the 'Geant,' although there was a thick haze underneath, through which we could occasionally see lights glimmering from the earth. We had attained a sufficient altitude to be only just able to hear noises from villages that we left beneath us, and were beginning to enjoy the delicious calm and repose peculiar to aerial ascents.

"There is, however, a talk about dinner, or rather supper, and night is now fast approaching. Every one eats with the best possible appetite. Hams, fowls and dessert only appear to disappear with an equal promptitude, and we quench our thirst with bordeaux and champagne. I remind our companions of the pigeons we brought with us, and which are hanging in a cage outside the railing. I knew there was no danger of their flying away, so fearlessly opened the cage. The three or four birds I had put in the car seemed struck with terror. They flew awkwardly towards the centre of our party, tumbling among the plates and dishes and under our feet. It was not a case of hunger with them, and I ought to have remembered that their feeding time was long since past. I replaced them in their cage.

"Meanwhile, the sun has left us for some time. Our longing gaze followed it behind the dark clouds in the horizon, whose edges it tipped with a glorious purple. Its last rays shone on us, and then came a bluish-grey twilight. Suddenly we are enveloped in a dense fog. We look around, above us. Everything has disappeared in the mist. The balloon itself is no longer visible. We can see nothing except the ropes which suspend us, and these are only visible for a few feet above our heads, when they lose themselves in the fog. We are alone with our wickerwork house in an unfathomable vault.

"We still ascend, however, through the compact and terrible fog, which is so solid-looking as to seem capable of being carved into forms with a knife. As we were without a moon, and had no light at all, in fact, we were unable to distinguish nicely the different shades of colour in these thick clouds. Now and then, when the clouds seemed to be lighter, they had a bluish tinge; but the thicker ones were dirty and muddy-looking. Dante must have seen some like these.

"Water trickled down our faces, hands, and clothes, and the ropes and sides of our car.

"The water did not fall in rain-drops or in flakes, as it sometimes does in the tropics; but we were as completely saturated by this heavy, penetrating mist as if we had been under a waterfall. We still continued to traverse these rainy regions. The thick fog which the balloon dislodged in forcing a passage closed immediately after it. At one moment I thought I felt something press against my cheek, which could only be compared to the points of a thousand needles, or to floating particles of ice. We were all of us too much absorbed with our situation to think of the hour or of the height to which we had attained. Suddenly the Prince of Wittgenstein, who was standing at my left hand, cried out under his breath—

"'Look at the balloon, sir! look at the balloon!'

"I raised my eyes, in company with several others, and shall never forget the magnificent sight which awaited them. I saw the balloon, for which I had been searching in vain a few minutes before. It had undergone a transformation. It looked now as if coated with silver, and floating in a pale phosphorescent glimmer. All the ropes and cords seemed to be of new, bright, and liquid silver, like mercury, caused by the mist which had rested on them becoming suddenly congealed. Two luminous arcs intervened between us, in a sea of mother-of-pearl and opal, the lower one being the colour of red ochre and the upper one orange. Both of them, blinding in their brilliancy, seemed about to embrace one another.

"'How far are they off?' thought I to myself. 'Can I touch them with my hand, or are they separated from me by an immense space?' We are not capable of forming ideas of perspective, floating as we are in the midst of such a glimmering splendour.

"Above and around us are nothing but thick fogs and enormous black clouds, whose ragged edges and backs are relieved by a pale silver coating. They undulate ceaselessly to and fro, and either usurp quietly the place of others, or disappear only to be superseded by more formidable ones. But the last ray of reflected light has died out, and we plunge into this chaos of dreadful forms. Monsters seem to wish to approach us, and to envelop us in their dark embraces. One of them, on my right hand, looks like a deformed human arm in a menacing attitude, writhing its jagged top like a blind serpent feeling its way. The vague monster has disappeared; but the momentary splendour being followed by the original gloom, we plunge once more into a darkness that can be felt.

"The water which had collected on the balloon during its ascent now began to take effect, and caused it to descend with such rapidity into the dark abyss that the ballast, which was immediately thrown overboard, was overtaken in its descent and fell on our heads again.

"I hear exclamations and voices near me. My companions are evidently agitated, and with good reason, too; for the lights which we could see a long way below us approach with terrible rapidity. We reached the earth rather quicker than we left it.

"Suddenly we feel a dreadful shock, followed by ominous crackings. The car has grounded. The 'Geant' has made its descent. But in what part of the habitable globe, and under what zone? At Meaux!"

To employ an expression of M. Nadar's it seems that these gentlemen never before experienced such a "knock-down blow."

After all these preparations, all this trouble, all the energy employed in the undertaking—sufficient, indeed, wherewith to attempt to cross the Atlantic—to "descend at Meaux!"

The 'Geant,' however, had its revenge. Its second ascent gave it this revenge. We shall be as brief as possible in relating this voyage; but the details are all so very interesting that we regret extremely our being unable to give more than extracts from the narrative.

Our travellers committed themselves again to the mercy of the air. The Emperor, following the example of a former King of France, took considerable interest in the construction of this aerial monster, and wished the aeronaut "Bon voyage" at starting. The passengers endeavoured to pass the night as comfortably as possible, having first instituted a four hours' watch, as on board ship.

The aerial vessel glided rapidly through the air. "We repeatedly," said Nadar, "passed over some manufacturing centre, whose lights were not yet extinguished. I either hailed them with my speaking-trumpet or rang our two bells. Sometimes we received a reply from below, in the shape of a shout, for, although we still had no moon, the night was occasionally clear enough for people to distinguish us; and sometimes we heard a peal of laughter from out of the atmosphere in which we were travelling. It was another party of aeronauts in a smaller balloon, who left at the same time as we did, and who would persist in keeping the 'Geant' company. We are passing over a small town; we hear the usual shouting and the report of a gun. Our first thoughts are—Was it loaded with shot or ball? The inhuman brute who fired will say, 'Certainly not;' but as balloons have often been damaged in this way, we may be confident there was more than powder in this one. It would be satisfactory, at any rate, if the name of the person could be ascertained who favoured us with this welcome. But it is rather late to make inquiries on this subject. It was between a quarter and half-past nine o'clock when this occurred. 'The sea!' cried Jules; 'look at the revolving lights of the lighthouses. There: one has just disappeared: it will flash out again in a moment!' But what is this? Before us, as far as our eyes can reach, we distinguish faint lights, which in this case are neither lamps nor torches. As we continue to draw nearer we get a better view of these numerous, violent, and smoking furnaces. Loud and ringing sounds strike on our ear at the same time. Am I right in my conjectures? Is this not that splendid country I love more than ever now? It must be Erquelines! And the dignified Custom-house official, had it been possible, would have added thereto 'Belgium!'

"We still continue to pass over fires, forges, tall chimneys, and coal mines at frequent intervals. Not long after we distinguish a large town on our right hand, which, by its size and brilliant lighting by gas, we recognise as Brussels. There could be no mistake, for close by, more modest in size and appearance, we see Catholic Malines. We have left it behind us.

"Onward! Onward! Behind us the fires fade gradually away, and disappear one after another. Before us nothing at present visible. We seemed to drift on for about one hundred or one hundred and fifty yards more. We cannot distinguish a single point in front of us on which to fix our gaze. But we still continue our course in silence.

"This mournful darkness, this endless shroud, in which we can discover neither rent nor spangle, still continues. Where are we? Over what strange country, possessing neither cities, towns, nor villages, are we hovering in the tomb-like silence of this interminable darkness? We seem, indeed, to have been carried by a puff of wind towards the west.

"But something seems to approach us. What are those pale rays of light which we can faintly see a long, long way before us—rays pale and soft, quite unlike those flaming fires we have left behind us? Surely these do not denote the presence of human activity! As we continue to advance, these pale flakes of light—resembling nothing so much in appearance as molten lead—which at first were scanty and isolated, gradually expand, and leave only narrow strips of darkness to divide them into fantastic shapes. By their help we discovered we were passing over the immense marshes of Holland, which extended to and lost themselves in the hazy horizon. On our right hand we hear a deep moan, still distant, but rapidly approaching every moment. It is undoubtedly the rushing of the wind. A fresh breeze for five minutes would bring us to the sea.

"We experienced another shock not less formidable than the first. The 'Geant' is trembling from its effects. The cable of our first anchor has just broken like a piece of thread. We could not hope for a better result. The violence of the wind which is carrying us along seems to be redoubled. A bump: another and another—then shock after shock.

"'The second dead men!'

"Our swift pace was shock after shock.

"'The anchor is lost,' cries Jules; 'we are all dead men!'

"This truth is too palpable to all of us to require expressing in so many words, for we are just commencing that furious, tearing course called 'trailing.'

"Our swift pace was considerably accelerated by the lower part of the balloon, which—limp, empty, and forming nearly a third of the whole—had been set free at the first shock, and flapped against the distended part, acting as a sail. The shocks continued to multiply so fast that it was impossible to count them. The car continued to rebound from these shocks to the height of five, ten, sometimes thirty, forty, and even fifty feet, for all the world like an India-rubber ball from the hands of an indefatigable player. Unfortunately, all our human freight, terror stricken and without advice, had crowded into one side of the car; and as this happened to be the side on which we invariably bumped, we experienced all the worst effects of the joltings.

"What a dizzy whirl! What a succession of breathless shocks! What a strain on both muscles and nerves! By the least negligence or slip, or by the loss of presence of mind for one moment, we should have been thrown out and dashed to atoms.

"Every collision tries our muscles and strains our wrists or our shoulders; and every rebound dashes us one against the other, constituting each individual a tormentor and victim at the same time. Our flight is so rapid that we can only distinguish an occasional glimpse of anything. Far, far in the distance we distinguish an isolated tree. We approach it like lightning, and we break it as though it were a straw.

"Two terrified horses, with manes and tails erect, endeavour to fly from us. But we consume distances; we leave them behind immediately. We skip over a flock of affrighted sheep in one of our bounds. But now comes the real danger.

"At this moment, when we were perfectly benumbed with fear, and had lost all power of articulation, we saw a locomotive, drawing two carriages, running along an embankment at right angles to our course. A few more revolutions of the wheels, and it will be all over with us, for we seem to be fated to meet with geometrical precision at one spot!

"What will happen?

"Travelling at our present hurricane pace, we shall undoubtedly lift up and overturn the machine and what it is drawing. But shall we not be crushed ourselves? A few paces still intervene between us and our foe, and we give vent to a shout of terror.

"It is heard, and the locomotive answers it by a whistle, then slackens its pace, and after seeming to hesitate an instant backs quickly and only just in time to give us a free passage, whilst the driver, waving his cap, salutes us with—

"'Look out for the wires!'

"The caution was well timed, for we had not noticed the four telegraph wires which we rapidly approached. We energetically ducked our heads on seeing them, but fortunately we escaped any more damage than having two or three of our ropes cut. These we continued to drag after us like the tail of a ragged comet, having the telegraph-wires and the posts which lately supported them attached to us."

After having been dragged thus for some time at the mercy of a hurricane which they ought to have been able to avoid, these aerial navigators at last got entangled in the outskirts of a wood near Rethem, in Hanover. A few broken arms and legs paid for their temerity in meddling with this monster, and one and all of the passengers have reason to be thankful that it will be unnecessary for us to proclaim their virtues and their fate in our next chapter.

Chapter X. The Necrology of Aeronautic

We will conclude this second part by giving a brief notice of some of those who, in the early days of aerostation, fell martyrs to their devotion to the new cause, and sometimes victims to their own want of foresight and their inexperience.

First among these is Pilatre des Roziers, with whose courage and ingenuity our readers are already familiar. After the passage of Blanchard from England over to France this hero, who was the first to trust himself to the wide space of the sky, resolved to undertake the return voyage from France to England—a more difficult feat, owing to the generally adverse character of the winds and currents. In vain did Roziers' friends attempt to make him understand the perils to which this enterprise must expose him; his only reply was that he had discovered a new balloon which united in itself all the necessary conditions of security, and would permit the voyager to remain an unusually long time in the air. He asked and obtained from government the sum of 40,000 livres, in order to construct his machine. It then became clear what sort of balloon he had contrived. He united in one machine the two modes previously made use of in aerostation. Underneath a balloon filled with hydrogen gas, he suspended a Montgolfiere, or a balloon filled with hot air from a fire. It is difficult to understand what was his precise object in making this combination, for his ideas seem to have been confused upon the subject. It is probable that, by the addition of a Montgolfiere, he wished to free himself from the necessity of having to throw over ballast when he wished to ascend and to let off this gas when he wished to descend. The fire of the Montgolfiere might, he probably supposed, be so regulated as to enable him to rise or fall at will.

This mixed system has been justly blamed. It was simply "putting fire beside powder," said Professor Charles to Roziers; but the latter would not listen, and depended for everything on his own intrepidity and scientific skill of which he had already given so many proofs. There were, perhaps, other reasons for his unyielding obstinacy. The court that had furnished him with the funds for the construction of the balloon pressed him, and he himself was most ambitious to equal the achievement of Blanchard, who was the first to cross the Channel, on the 7th of January, 1785.

The fact was that at this time the prevailing fear in France was, that Great Britain should bear off all the honours and profits of aerostation before any of these had been won by France. It was thus that with an untried machine, and under conditions the most unfavourable for his enterprise, Roziers prepared to risk his life in this undertaking, which was equally dangerous and useless.

The double balloon was alternately inflated and emptied. While under cover it was assailed by the rats that gnawed holes in it, and when brought out of its place it was exposed to the tempests, so that the longer the experiment was delayed, the worse chance there was of getting through it successfully. At length Roziers went to Boulogne, and announced the day of his departure; but, as if by a special Providence, his attempt was delayed by unfavourable weather. For many weeks in succession the little trial balloons thrown up to show the course of the wind were driven back upon the shores of France. During all these trials the impatient Roziers continued to chafe and torment himself.

At last, on the 13th and 14th of June, 1785, the Aero-Montgolfiere remained inflated, waiting a favourable moment for departure. On the 15th at four in the morning, a little pilot balloon that had been thrown up fell back on the spot from which it had been thrown free, thus showing that there was no wind. Seven hours later Roziers, accompanied by his brother Romain, one of the constructors of the balloon, appeared in the gallery. A nobleman present threw a purse of 200 louis into the car, and was preparing to follow it and join in the adventure. Roziers forbade him to enter, gently but firmly.

"The experiment is too unsafe," he said, "for me to expose to danger the life of another."

"Finally," says a narrative of the time, "the Aero-Montgolfiere rose in an imposing manner. The sound of cannon signalised the departure, the voyagers saluted the crowd, who responded with loud shouts. The balloon advanced until it began to traverse the sea, and every one with eyes fixed upon the fragile machine, regarded it with fear. It had traversed upwards of a league of its journey, and had reached the height of 700 feet above sea level, when a wind from the west drove it back toward the shore, after having been twenty-seven minutes in the air.

"At this moment the crowd beneath perceived that the voyagers were showing signs of alarm. They seemed suddenly to lower the grating of the Montgolfiere. But it was too late. A violet flame appeared at the top of the balloon, then spread over the whole globe, and enveloped the Montgolfiere and the voyagers. "The unfortunate men were suddenly precipitated from the clouds to the earth, in front of the Tour de Croy, upwards of a league from Boulogne, and 300 feet from the sea beach.

"The dead body of Roziers was found burnt in the gallery, many of the bones being broken. His brother was still breathing, but he was not able to speak, and in a few minutes he expired."

De Maisonfort, who, against his own will, was left on the earth, was witness of this sad event. He has given the following explanation of it: —

"Some minutes after their departure the voyagers were assailed by contrary winds, which drove them back again upon the land. It is probable that then, in order to descend and seek a more favourable current of air, which would take them out again to sea, Roziers opened the valve of the gas balloon; but the cord attached to this valve was very long, it worked with difficulty, and the friction which it occasioned tore the valve. The stuff of the balloon, which had suffered much from many preliminary attempts, and from other causes, was torn to the extent of several yards, and the valve fell down inside the balloon, which at once emptied itself."

According to this narrative, there was no conflagration of the gas in the middle of the atmosphere, nor is it stated precisely whether the grating of the Montgolfiere was lighted.

Maisonfort ran to the spot when the travellers fell, found them covered with the cloth of the balloon, and occupying the same positions which they had taken up on departing.

By a sad chance, that seems like irony, they were thrown down only a few paces from the monument which marks the spot where Blanchard descended. At the present day Frenchmen going to England via Calais do not fail to visit at the forest of Guines the monument consecrated to the expedition of Blanchard. A few paces from this monument the cicerone will point out with his finger the spot where his rivals expired.

"Such was the end of the first of aeronauts, and the most courageous of men," says a contemporaneous historian. "He died a martyr to honour and to zeal. His kindness, amiability, and modesty endeared him to all who knew him. She who was dearest to him — a young English lady, who boarded at a convent at Boulogne, and whom he had first met only a few days prior to his last ascent — could not support the news of his death. Horrible convulsions seized her and she expired, it is said, eight days after the dreadful catastrophe. Roziers died at the age of twenty-eight and a half years."

Olivari perished at Orleans on the 25th of November, 1802. He had ascended in a Montgolfiere made of paper, strengthened only by some bands of cloth. His car, made of osiers, and loaded with combustible matter, was suspended below the grating; and when at a great elevation it became the prey of the flames. The aeronaut, thus deprived of his support, fell, at the distance of a league from the spot from which he had risen.

Mosment made his last ascent at Lille on the 7th of April, 1806. His balloon was made of silk, and was filled with hydrogen gas. Ten minutes after his departure he threw into the air a parachute with which he had provided himself. It is supposed that the oscillations consequent on the throwing off of the parachute were the cause of they aeronaut's fall. Some pretend that Mosment had foretold his death, and that it was caused by a willful carelessness. However this may be, the balloon continued its flight alone, and the body of the aeronaut was found partly buried in the sand of the fosse which surrounds the town.

Bittorff made a great many successful ascents. He never used any machine but the Montgolfiere. At Manheim, on the 17th of July, the day of his death his balloon, which was of paper, sixteen metres in diameter, and twenty in height, took fire in the air, and the aeronaut was thrown down upon the town. His fall was mortal.

Harris, an old officer of the English navy, together with another English aeronaut, named Graham, had made a great many ascents.

He conceived the idea of constructing a balloon upon an original plan; but his alterations do not seem to have been improvements. In May, 1824, he attempted an ascent from London, which had much apparent success, but which terminated fatally. When at a great elevation, it seems, the aeronaut, wishing to descend, opened the valve. It had not been well constructed, and after being opened it would not close again. The consequent loss of gas brought the balloon down with great force. Harris lost his life with the fall; but the young lady who had accompanied him received only a trifling wound.

Sadler, a celebrated English aeronaut, who, in one of his many experiments, had crossed the Irish Channel between Dublin and Holyhead, lost his life miserably near Bolton, on the 28th of September, 1824. Deprived of his ballast, in consequence of his long sojourn in the air, and forced at last to descend, at a late hour, upon a number of high buildings, the wind drove him violently against a chimney. The force of the shock threw him out of his car, and he fell to the earth and died. His prudence and knowledge were unquestionable, and his death is to be ascribed alone to accident. It was an aerial shipwreck.

Cocking had gone up twice in Mr. Green's balloon as a simple amateur. He took it into his head to go up a third time. He wished to attempt a descent in a parachute of his own construction, which he believed was vastly superior to the ordinary one. He altered the form altogether, though that form had been proved to be satisfactory. In place of a concave surface, supporting itself on a volume of air, Cocking used an inverted cone, of an elaborate construction, which, instead of supporting him in the air, only accelerated his fall. Unhappily, Green participated in this experiment. The two made an ascent from Vauxhall, on the 27th of September, 1836, Green having suspended Cocking's wretched contrivance from the car of his balloon. Cocking held on by a rope, and at the height of from 1,000 to 1,200 feet the amateur, with his patent parachute, were thrown off from the balloon. A moment afterwards Green was soaring away safely in his machine, but Cocking was launched into eternity.

"The descent was so rapid," says one who witnessed it, "that the mean rate of the fall was not less than twenty yards a second. In less than a minute and a half the unfortunate aeronaut was thrown to the earth, and killed by the fall."

Madame Blanchard, thinking to improve upon Garnerin, who had decorated the balloon which ascended in celebration of the coronation of Napoleon I. with coloured lights, fixed fireworks instead to hers. A wire rope ten yards long was suspended to her car; at the bottom of this wire rope was suspended a broad disc of wood, around which the fireworks were ranged. These consisted of Bengal and coloured lights. On the 6th of July, 1819, there was a great fete at Tivoli, and a multitude had assembled around the balloon of Madame Blanchard. Cannon gave the signal of departure, and soon the fireworks began to show themselves. The balloon rose splendidly, to the sound of music and the shoutings of the people. A rain of gold and thousands of stars fell from the car as it ascended. A moment of calm succeeded, and then to the eyes of the spectators, still fixed on the balloon, an unexpected light appeared. This light did not come from under the balloon, where the crown of fireworks was already extinguished, but shone in the car itself. It was evident that the lady aeronaut, although now so high above the spectators, was busy about something. The light increased, then disappeared suddenly; then appeared again, and showed itself finally at the summit of the balloon, in the form of an immense jet of gas. The gas with which the balloon was inflated had taken fire, and the terrible glare which the light threw around was perceived from the boulevards, and all the Quartier Montmartre.

It was at this moment—a frightful one for those who perceived what had taken place—that a general sentiment of satisfaction and admiration among the spectators found vent in cries of "Brava! Vive Madame Blanchard!" &c. The people thought the lady was giving them an unexpected treat. Meantime, by the light of the flame, the balloon was seen gradually to descend. It disappeared when it reached the houses, like a passing meteor, or a train of fire which a blast of wind suddenly extinguishes. A number of workmen and other persons, who had perceived that some accident had taken place, ran in the direction in which the balloon appeared to descend. They arrived at a house in the Rue de Provence. On the roof of this house the balloon had fallen, and the unfortunate Madame

Blanchard, thrown out of the car by the shock, was killed by her fall to the earth.

This news spread rapidly from Tivoli, where it occasioned a stupefying surprise. It was the first time that a fall of the kind had taken place from the sky at Paris. Fireworks were from this time discontinued, the fete came to an end, and a subscription was rapidly organised, producing some thousands of francs, which shortly afterwards were employed in erecting a monument to the lady, which is now to be seen in the cemetery of Pere-la-Chaise.

Madame Blanchard had wished to surpass the ordinary spectacle of an aerial ascent; she had really prepared a SURPRISE for the spectators. She had prepared and she took with her a small parachute of about two yards in diameter. After the extinction of the crown or star of fireworks, she intended to throw this little parachute loose; and as it was terminated by another supply of fireworks, it was supposed that the effect would be as beautiful as surprising.

The unhappy lady was small in stature, and very light, and unfortunately made use of a very small balloon. That of the 6th of July, 1819, was only seven metres in diameter; and to make it ascend with the weight it carried it had to be filled to the neck with inflammable air. In quitting the earth some of this gas escaped, and rising above the balloon, formed a train like one of powder, which would certainly flash into a blaze the moment it came in contact with the fire. But on this day it was she who with her own hand fired this train. At the moment when, detaching the little parachute from her car, she took the light for the fireworks in her other hand, she crossed this train with the light and set it on fire. Then the brave woman, throwing away the parachute and the match, strove to close the mouth of the balloon, and to stifle the fire. These efforts being unavailing, Madame Blanchard was distinctly seen to sit down in her car and await her fate.

The burning of the hydrogen lasted several minutes, during which time the balloon gradually descended. Had it not been that it struck on the roof of the house Madame Blanchard would have been saved. At the moment of the shock she was heard to cry out, "A moi." These were her last words. The car, going along the roof of the house, was caught by an iron bar and overturned, and the lady was thrown head foremost upon the pavement.

When she reached the ground she immediately expired. Her head and shoulders were slightly burnt, otherwise she exhibited no marks of the fire which had destroyed the balloon.

PART III. Scientific Experiments — Applications of Ballooning.

Chapter I. Experiments of Robertson, Lhoest, Saccarof, &c.

Robertson is regarded by many as a sort of mountebank; yet such men as Arago have put themselves to the trouble of examining the aerostatic feats of this aeronaut, and of examining the results of his observations.

"The savant Robertson," says Arago, "performed at Hamburg on the 18th of July, 1803, with his countryman, Lhoest, the first aeronautic voyage from which science has been able to draw useful deductions. The two aeronauts remained suspended in the air during five hours, and came down near Hanover, twenty-five leagues from the spot from which they set off."

The first time that Robertson appears in the annals of aerostation is in 1802, on the occasion of the sale of the balloon used at the battle of Fleurus, of which mention will be made in the chapter on military aerostation. But three years previously he had been instructed to make a balloon of an original form, which should ascend in honour of the Turkish ambassador at the garden of Tivoli. The fete was completely successful. Turks, Chinese, Persians, and Bedouins will always be welcome, as on this occasion, at Paris, appearing as they do only at rare intervals, and for a short time.

The fete took place on the 2nd of July. Robertson presented himself at the house of Esseid-Ali, to obtain his autograph. The Turkish ambassador willingly granted the request, and wrote his name in letters, each of which was two inches in height, on a sheet of paper. He then offered the aeronaut coffee and comfits, and promised to be present to witness the balloon ascent. His name was painted in large characters on a balloon fifteen feet in diameter, and on the form of which was the figure of a crescent. The experiment delighted the ambassador, and was well received by the public.

Jacques Garnerin, when he came to make his debut as an aeronaut, made an attempt with the parachute, the following August, at the garden of the Hotel de Biron. The ambassador was asked to honour the fete, but he declined, saying that he had "made up his mind that man was not intended for flying—Mahomet had not so willed it."

Of one of Robertson's more interesting ascents he himself has left us the following sketch:—

"I rose in the balloon at nine a.m., accompanied by my fellow-student and countryman, M. Lhoest. We had 140 lbs. of ballast. The barometer marked twenty-eight inches; the thermometer sixteen degrees Reaumur. In spite of some slight wind from the north-west, the balloon mounted so perpendicularly that in all the streets each of the spectators believed we were mounting straight up above his head. In order to quicken our ascent I discharged a parachute made of silk, and weighted in a way to prevent oscillations. The parachute descended at the rate of two feet per second, and its descent was uniform. From the moment when the barometer began to sink we became very careful of our ballast, as we wished to test from experience the different temperatures through which we were about to pass.

"At 10.15, the barometer was at nineteen inches, and the thermometer at three above zero. We now felt all the inconvenience of an extremely rarefied atmosphere coming upon us, and we commenced to arrange some experiments in atmospheric electricity. Our first attempts did not succeed. We threw over part of our ballast, and mounted up till the cold and the rarefaction of the air became very troublesome. During our experiments we experienced an illness throughout our whole system. Buzzing in the cars commenced, and went on increasing. The pain we felt was like that which one feels when he plunges his head in water. Our chests seemed to be dilated, and failed in elasticity. My pulse was quickened, M. Lhoest's became slower; he had, like me, swelled lips and bleeding eyes; the veins seemed to come out more strongly on the hands. The blood ran to the head, and occasioned a feeling as if our hats were too tight. The thermometer continued to descend, and, as we ascended, our illness increased, and we could with difficulty keep awake. Fearing that my travelling companion might go to sleep, I attached a cord to my thigh and to his, and we held the extremities of the cord in our hands. Thus trammelled, we had to commence the experiments which I had proposed to make.

"At this elevation, the glass, the brimstone, and the Spanish wax were not electrified in a manner to show any signs under friction—at least, I obtained no electricity from the conductors or the electrometer.

"I had in my car a voltaic pile, consisting of sixty couples—silver and zinc. It worked very well on the occasion of our departure from the earth, and gave, without the condenser, one degree to the electrometer. At our great elevation, the pile gave only five-sixths of a degree to the same electrometer. The galvanic flame seemed more active at this elevation than on the earth.

"I took two birds with me on coming into the balloon—one of these was now dead, the other appeared stupefied. After having placed it upon the brink of the gondola, I tried to frighten it to make it take to flight. It moved its wings, but did not leave the spot; then I left it to itself, and it fell perpendicularly and with great rapidity. Birds are certainly not able to maintain themselves at such elevations.

"It is notable that the atmosphere, which was of a perfect purity near the earth, was grey and misty above our heads, and the beautiful blue sky seen from the surface did not exist for us, although the weather was calm and serene, and the day the most beautiful that could be. The sun did not seem dazzling to us, and its heat was diminished owing to our elevation.

"At half-past eleven, the balloon was no longer visible from Hamburg. The heavens were so pure beneath us that everything was distinctly seen by us, though very much diminished by distance. At 11.40, the town of Hamburg seemed only a red point in our eyes; the Elbe looked like a straight ribbon. I wished to make use of an opera-glass, but what surprised me was that when I lifted it up it was so cold that I had to wrap my handkerchief around it to enable me to hold it.

"Not being able to support our position any longer, we descended, after having used up much gas and ballast. Our descent caused that degree of terror among the inhabitants which the size of our balloon was calculated to inspire in a country where such machines had never before been seen. We descended above a poor village called Radenburg, a place amid the heaths of Hanover. Our appearance caused great alarm, and even the beasts of the field fled from us.

"While our balloon rapidly approached the earth, we waved our hats and flags, and shouted to the inhabitants, but our voices only increased their terror. The villagers rushed away with cries of terror, leaving their herds, whose bellowings increased the general alarm. When the balloon touched the ground, every man had shut himself up in his own house. Having appealed in vain, and fearing that the villagers might do us some injury, we resolved to re-ascend.

"In making this second ascent, we threw over all our ballast; but in this we were imprudent, for after sailing about at a great height, and having lost much gas, I perceived that our descent would be very rapid, and to provide against accident, I gathered together all the instruments, the bread, the ropes, and even such money as we had with us, and placed them in three sacks, to which I attached a rope of a hundred feet in length. This precaution saved us a shock. The weight, amounting to thirty pounds, reached the ground before us, and the balloon, thus lightened, came softly to the ground between Wichtenbech and Hanover, after having run twenty-five leagues in five and a half hours."

After this ascent Robertson became acquainted with some savants of Hamburg, and amongst others with Professor Pfaff, who was interested in aerial travelling as a means of settling certain meteorological problems. Some days after Robertson's ascent, the professor wrote to him —

"You speak of a certain height at which the hydrogen gas will find itself in equilibrium in the air of the atmosphere. I believe that this height is the extremity of the atmosphere itself; for as the gas has an elasticity much greater than that of the air, it will go on dilating as it mounts into the higher regions of the atmosphere, and its specific weight will diminish as the weight of atmospheric air diminishes; and it will not cease to mount until it rises above the atmosphere itself, if two conditions be completely fulfilled—1, the condition that the gas may be allowed to dilate without leaving the balloon as it rises; 2, the condition that the gas shall not be allowed to mix at all with the atmospheric air."

Another ascent was arranged for the 14th of August, in which Robertson was to be accompanied by the professor, but the latter, yielding to the entreaties of his family, did not go. "I went up with my friend Lhoest," says Robertson, "at forty-two minutes past twelve midday. In a minute or two we rose up between two masses of cloud, which seemed to open up and offer us a passage. The upper surfaces of these clouds are not uniformly level, like the under sides seen from the earth, but they are of a conical or pyramidal shape. These imposing masses seem to precipitate themselves upon the earth, as if to engulf it, but this optical illusion was due to the apparent immobility of the balloon, which at the moment was rising at the rate of about twenty feet per second.

"The fear of losing the view of the Baltic, which we perceived between the clouds at intervals, obliged us to renounce the project of rising as high as on the last occasion. The barometer was at fifteen inches, and the thermometer one degree below zero, when I let off two pigeons.

"One descended in a diagonal direction, its wings half open but not moving, with a swiftness which seemed that of a fall. The other flew for an instant, and then placed itself upon the car, and did not wish to quit us. Acting on the hint of Dr. Reimarus, I tried the same experiment with butterflies, but the air was too much rarefied for them; they attempted in vain to raise themselves by their wings, but they did not forsake the car.

"The wind continuing to carry me towards the sea, I resolved to bring my observations to an end. I effected my descent in a meadow, near the village of Rehorst, in Holstein, after having run sixteen leagues from France in sixty-five minutes."

At the commencement of the year 1804, Laplace, at the Institute, proposed to take advantage of the means offered by aerostation to verify at great heights certain scientific points—as, for example, those which concern magnetism. This proposition was made at a favourable time, and was, so far, carried out in the best possible way. The aeronauts who were appointed to carry out the expedition were Biot and Gay-Lussac, the most enthusiastic aeronauts of the period.

The following is their report:—

"We observed the animals we had with us at all the different heights, and they did not appear to suffer in any manner. For ourselves, we perceived no effect any more then a quickening of the pulse. At 10,000 feet above the ground we set a little green-finch at liberty. He flew out at once, but immediately returning, settled upon our cordage; afterwards, setting out again, he flew to the earth, describing a very tortuous line in his passage. We followed him with our eyes till he was lost in the clouds. A pigeon, which we set free at the same elevation, presented a very curious spectacle. Placed at liberty on the edge of the car, he remained at rest for a number of instants, as if measuring the length of his flight; then he launched himself into space, flying about irregularly, as if to try his wings. Afterwards he began his descent regularly, sweeping round and round in great circles, ever reaching lower, until he also was lost in the clouds."

As to the voyagers themselves, this is how they speak of their situation at the height of 3,000 yards:—

"About this elevation we observed our animals. They did not appear to suffer from the rarity of the air, yet the barometer was at twenty inches eight lines.. We were much surprised that we did not suffer from the cold; on the contrary, the sun warmed us much. We had thrown aside the gloves which had been put on board, and which were of no use to us. Our pulses were very quick; that of M. Gay-Lussac, which is 62 in the minute on ordinary occasions, now gave 80; and mine, which is ordinarily 89, gave 111. This acceleration was felt by both of us in nearly the same proportion. Nevertheless, our respiration was in no way interfered with, we experienced no illness, and our situation seemed to us extremely agreeable."

The following is their report to the Galvanic Society—

"We have known for a long time that no animal can with safety pass into an atmosphere much more dense or much more rare than that to which it has been accustomed. In the first case it suffers from the outer air, which presses upon it severely; in the second case there are liquids or fluids in the animal's body which, being less pressed against than they should be, become dilated, and press against their coverings or channels. In both cases the symptoms are nearly the same — pain, general illness, buzzing in the ears, and even haemorrhage. The experience of the diving-bell has long made us familiar with what aeronauts suffer. Our colleague (Robertson), and his companion, have experienced these effects in great intensity. They had swelled lips, their eyes bled, their veins were dilated, and, what is very remarkable, they both preserved a brown or red tinge which astonished those that had seen them before they made the ascent. This distension of the blood-vessels would necessarily produce an inconvenience and a difficulty in the muscular action."

Chapter II. Ascent of M. Gay-Lussac Alone — Excursions of MM. Barral and Bixio.

Respecting this ascent, Arago states that M. Gay-Lussac has reduced to their proper value the narratives of the physical pains which aeronauts say they suffer in lofty aerial ascents.

M. Gay-Lussac says: — "Having arrived at the most elevated point of my ascent, 21,000 feet above sea level, my respiration was rendered sensibly difficult, but I was far from experiencing any illness of a kind to make me descend. My pulse and my breathing were very quick; breathing very frequently in an extremely dry atmosphere, I should not have been surprised if my throat had been so dry as to make it painful to swallow bread."

After having finished his observations, which referred chiefly to the magnetic needle, with all the tranquillity of a doctor in his study, Gay-Lussac descended to the earth between Rouen and Dieppe, eighty leagues from Paris.

After the names of Robertson, Gay-Lussac, and Biot, science has registered those of Barral and Bixio, two men whose aeronautic achievements have enriched meteorology with more important discoveries, perhaps, than any we have yet mentioned.

These gentlemen had conceived the project of rising by means of a balloon to a great height, in order to study, with the assistance of the very best instruments in use in their day, a multitude of phenomena then imperfectly known. The subjects to which they were specially to direct their attention, were the law of the decrease of temperature in progress upwards, the discovery of whether the chemical composition of the atmosphere is the same throughout all its parts, the comparison of the strength of the solar rays in the higher regions of the atmosphere and on the surface of the earth, the ascertaining whether the light reflected and transmitted by the clouds is or is not polarised, &c.

All the preparations having been made in the garden of the Observatory at Paris, the ascent took place on the 29th of June, 1850, at 10.27 a.m., the balloon being filled with hydrogen gas. The first ascent was a signal failure. It was found that the weather being bad, the envelope of the balloon was torn in several places, and had to be mended in all haste. Immediately preceding the moment of ascent, a torrent of rain fell. But the voyagers were determined to ascend. They placed themselves in the car, and, when thrown off from the fastenings, they rose through the air with the speed of an arrow. The height to which the balloon reached made it suddenly dilate, and the network, which was much too small, was stretched to the utmost. The balloon was forced down upon them by the dilation, and one of them, in the endeavour to work the valve, made a rent in the lower part of the globe, from which the gas escaping almost over the heads of the travellers, nearly choked them. The escape of the gas had the usual result—the balloon descended rapidly, and fell in a vineyard near Lugny, where they were found by the peasants holding on to the trees by their legs and arms, and thus attempting to stop the horizontal advance of the car. They had risen to the height of over 17,000 feet, and they had descended from this height in from four to five minutes.

For all practical purposes, the ascent was a failure, and the aeronauts immediately commenced preparations for a new voyage, which took place a month afterwards. They rose to very great altitudes, but experienced no illness from the rarefied air. M. Bixio did not feel the sharp pains in the ears from which he had suffered on the former occasion. They passed through a mass of cloud 15,000 feet in thickness, and they had not yet passed quite through it, when at the height of over 21,000 feet from the ground, they began to descend, their descent being caused by a rent in the envelope of the balloon, from which the gas escaped. They might, in throwing out the last of their ballast, have, perhaps, prolonged for a little their sojourn in space, but the circumstances in which they were placed did not permit them to make many more scientific observations than those they had made, and thus they were obliged to submit to their fate. When they had reached their greatest height, there seemed to open up in the midst of the vaporous mass a brilliant space, from which they could see the blue of heaven. The polariscope, directed towards this region, showed an internal polarisation, but, when pointed to the side where the mist still prevailed, there was no polarisation.

An optical phenomenon of a remarkable kind was witnessed when the voyagers had attained their highest point. They saw the sun through the upper mists, looking quite white, as if shorn of its strength; and, at the same time, below the horizontal plane, below their horizon, and at an angular distance from the plane equal to that of the sun above it, they saw a second sun, which resembled the reflection of the actual sun in a sheet of water. It is natural to suppose that the second sun was formed by the reflection of the sun's rays upon the horizontal faces of the ice crystals floating in this high cloud.

Chapter III. Ascents of the Mssrs. Welsh, Glaisher and Coxwell.

The most recent balloon ascents in England deserving attention have been undertaken for scientific objects, and in this country, more than in any other, it may be said that the conquest of the air has been made to serve a practical end.

In July, 1852, the Committee of the Kew Observatory resolved to undertake a number of balloon voyages. This resolution was approved of by the British Association for the Advancement of Science, and the necessary instruments for making a number of meteorological observations were prepared. The balloon employed was that of Mr. Green, who was accompanied in his ascents by Mr. Welsh. The greatest height to which Mr. Welsh rose was on the fourth ascent which took place on the 10th of November, 1852. The balloon rose 22,930 feet, and the lowest temperature observed was 26 degrees below zero.

It is to Mr. Glaisher and Mr. Coxwell, however, that the highest honours of scientific aerostation belong. The ascents made by these gentlemen — Mr. Glaisher being the scientific observer, and Mr. Coxwell the practical aeronaut — have become matters of history. Not only did they, in the course of a large number of ascents undertaken under the auspices of the British Association, succeed in gathering much valuable meteorological information, but they reached a greater height than that ever gained on any previous or subsequent occasion, and penetrated into that distant region of the skies in which it has been satisfactorily proved that no life can be long maintained. It was on the 5th of September, 1862, that Mr. Glaisher and Mr. Coxwell made the famous ascent in which they reached the greatest height ever attained by an aeronaut, and were so nearly sacrificed to their unselfish daring. Mr. Glaisher has given an admirable account of this ascent, which took place from Wolverhampton. He says: — "Our ascent had been delayed, owing to the unfavourable state of the weather. It commenced at three minutes past one p.m., the temperature of the air being 59 degrees, and the dew-point 48 degrees. At the height of one mile the temperature was 41 degrees and the dew-point 38 degrees. Shortly after wards clouds were entered of about 1,100 feet in thickness. Upon emerging from them at seventeen minutes past one, I tried to take a view of their surface with the camera, but the balloon was ascending too rapidly and spiraling too quickly to allow me to do so. The height of two miles was reached at twenty-one minutes past one. The temperature of the air had fallen to 32 degrees and the dew-point to 26 degrees. The third mile was passed at twenty-eight minutes past one, with an air temperature of 18 degrees, and a dew-point of 13 degrees. The fourth mile was passed at thirty-nine minutes past one, with an air temperature of 8 degrees, and a dew-point of minus 6 degrees and the fifth mile about ten minutes later, with an air temperature minus 5 degrees, and a dew-point minus 36 degrees.

"Up to this time I had experienced no particular inconvenience. When at the height of 26,000 feet I could not see the fine column of the mercury in the tube; then the fine divisions on the scale of the instrument became invisible. At that time I asked Mr. Coxwell to help me to read the instruments, as I experienced a difficulty in seeing them. In consequence of the rotary motion of the balloon, which had continued without ceasing since the earth was left, the valve line had become twisted, and he had to leave the car, and to mount into the ring above to adjust it. At that time I had no suspicion of other than temporary inconvenience in seeing. Shortly afterwards I laid my arm upon the table, possessed of its full vigour; but directly after, being desirous of using it, I found it powerless. It must have lost its power momentarily. I then tried to move the other arm, but found it powerless also. I next tried to shake myself, and succeeded in shaking my body. I seemed to have no legs. I could only shake my body. I then looked at the barometer, and whilst I was doing so my head fell on my left shoulder. I struggled, and shook my body again, but could not move my arms. I got my head upright, but for an instant only, when it fell on my right shoulder; and then I fell backwards, my back resting against the side of the car, and my head on its edge. In that position my eyes were directed towards Mr. Coxwell in the ring. When I shook my body I seemed to have full power over the muscles of the back, and considerable power over those of the neck, but none over my limbs. As in the case of the arms, all muscular power was lost in an instant from my back and neck. I dimly saw Mr. Coxwell in the ring, and endeavoured to speak, but could not do so; when in an instant intense black darkness came over me, and the optic nerve lost power suddenly. I was still conscious, with as active a brain as whilst writing this. I thought I had been seized with asphyxia, and that I should experience no more, as death would come unless we speedily descended. Other thoughts were actively entering my mind when I suddenly became unconscious, as though going to sleep. I could not tell anything about the sense of hearing: the perfect stillness of the regions six miles from the earth—and at that time we were between six and seven miles high—is such that no sound reaches the ear. My last observation was made at 29,000 feet, about fifty-four minutes past one. I suppose two or three minutes elapsed between my eyes becoming insensible to seeing the fine divisions and fifty-four minutes past one, and that other two or three minutes elapsed before I became unconscious; therefore I think that took place about fifty-six or fifty-seven minutes past one. Whilst powerless I heard the words 'temperature,' and 'observation,' and I knew Mr. Coxwell was in the car, speaking to me, and endeavouring to rouse me; and therefore consciousness and hearing had returned. I then heard him speak more emphatically, but I could not speak or move. Then I heard him say, 'Do try; now do!' Then I saw the instruments dimly, next Mr. Coxwell, and very shortly I saw clearly. I rose in my seat and looked round, as though waking from sleep, and said to Mr. Coxwell, 'I have been insensible.' He said, 'Yes; and I too, very nearly.' I then drew up my legs, which had been extended out before me, and took a pencil in my hand to note my observations. Mr. Coxwell informed me that he had lost the use of his hands, which were black, and I poured brandy over them. I resumed my observations at seven minutes past two. I suppose three or four minutes were occupied from the time of my hearing the words 'temperature' and 'observation,' until I began to observe. If so, then returning consciousness came at four minutes past two, and that gives about seven minutes of total insensibility. Mr. Coxwell told me that in coming from the ring he thought for a moment that I had laid back to rest myself; that he spoke to me without eliciting a reply; that he then noticed that my legs projected, and my arms hung down by my side. That my countenance was serene and placid, without earnestness or anxiety, he had noticed before going into the ring. It then struck him that I was insensible. He wished then to approach me, but could not, and he felt insensibility coming over himself. He became anxious to open the valve, but, in consequence of having lost the use of his hands, he could not; and ultimately he did so by seizing the cord with his teeth and dipping his head two or three times. No inconvenience followed our insensibility. When we dropped it was in a country where no accommodation of any kind could be obtained, so that we had to walk between seven and eight miles. At the time of ceasing our observations the ascent was at the rate of 1,000 feet per minute, and on resuming observations the descent was at the rate of 2,000 feet per minute. These two positions must be connected, having relation to the interval of time which elapsed between them; and they can scarcely be connected at a point less than 36,000 or 37,000 feet high. Again, a very delicate minimum thermometer was found to read minus 12 degrees, and that reading would indicate an elevation exceeding 36,000 feet. There cannot be any doubt that the balloon attained the great height of seven miles—the greatest ever reached. In this ascent six pigeons were taken up. One was thrown out at three miles. It extended its wings, and dropped like a piece of paper. A second at four miles, and it flew with vigour. A third between four and five miles, and it fell downwards. A fourth was thrown out at four miles in descending, and it alighted on the top of the balloon. Two were brought to the ground. One was dead, and the other was ill, but recovered so as to fly away in a quarter of an hour."

The results gathered by Mr. Glaisher from his numerous ascents are very interesting. He found that in no instance did the temperature of the air decrease uniformly with the increase of height. In fact, the decrease in the first mile is double that in the second, and nearly four times as great as the change of temperature in the fifth mile. The distribution of aqueous vapour in the air is no less remarkable. The temperature of the dew-point on leaving the earth decreases less rapidly than the temperature of the air; so that the difference between the two temperatures becomes less and less, till the vapour or cloud plane is reached, when they are usually together, and always most nearly approach each other, and that point is usually at about the height of one mile. On leaving the upper surface of cloud, the dew point decreases more rapidly than the air, and at extremely high situations the difference between the two temperatures is wonderfully great, indicating an extraordinary degree of dryness, and an almost entire absence of water. Under these circumstances, the presence of cirrus clouds far above this dry region, apparently as much above as when viewed from the earth, is very remarkable, and leads to the conclusion that they are not composed of water.

In the propagation of sound, M. Glaisher made many curious experiments. In one ascent (July 17th) he found, when at a distance of 11,800 feet above the earth, that a band was heard; at a height of 22,000 feet, a clap of thunder was heard; and at a height of 10,070 feet, the report of a gun was heard. On one occasion, he heard the dull hum of London at a height of 9,000 feet above the city, and on another occasion, the shouting of many thousands of persons could not be heard at the height of 4,000 feet.

Chapter IV. Balloons Made Useful in Warfare.

> Wars of the French Republic — Company of "Ballooneers" —
> Battle of Fleurus — The Balloons of Egypt — Napoleon —
> Modern Services War in Italy — War in America — Conclusion.

We will conclude our work with a glance at aerostation as applied to warfare. Scarcely had the first ascents astonished the world, than the more adventurous spirits began to use the new discovery for a thousand purposes directly useful to man. The first point of view in which aerostation was regarded, was in that of its practical utility If one refers to the pre-occupations of the time — to the great events then occurring in the history of France, one will easily understand that the Committee of Public Safety soon thought of employing balloons in the observation of the forces and the movements of hostile troops. In 1794, the idea was practically carried out, and the French armies were provided with two companies of aeronauts. The command of one of these companies was given to Captain Coutelle, a young physicist of great talent, who rendered memorable services at the battle of Fleurus. The balloons were not thrown free, but were retained attached by means of long cords. In this way they took up, so to speak, aerial posts of observation. Placed in his car, the captain transmitted his instructions to his men below by means of coloured flags. Coutelle has left us a lively narrative of certain incidents connected with one of the grand days of the old Republic. He had been commissioned by the Committee of Public Safety to go to Maubeuge, where Jourdan's army was encamped, and to offer him the use of his balloon. The representative to whom the young doctor presented his commission, knew nothing about balloons, and not being able to understand the order of the Committee of Public Safety, it suddenly dawned upon him that Coutelle, with his trumpery forgery about balloons, was nothing else than a spy, and he was about to have him shot. The genuineness of the order from the Committee, however, was proved, and Coutelle's case was listened to.

"The army was at Beaumont," says Coutelle, "and the enemy, placed at a distance of only three miles, could attack at any moment. The general told me this fact, and engaged me to return and communicate it to the Committee. This I did. The Commission then felt the necessity of making an experiment with a balloon that could raise two persons, and the minister placed at my service the garden and the little mansion of Meudon. Many of the members of the Commission came to witness the first ascent of a balloon held in hand, like a kite, by means of two cords. The Commissioners ordered me to place myself in the car, and instructed me as to a number of signals which I must repeat, and observations which I must make. I raised myself to the full length of the cord, a height of 1,500 feet, and at this height, with the help of a glass, I could distinctly see the seven bends of the river Seine. On returning to the earth, I received the compliments of the Commission.

"Arrived at Maubeuge, my first care was to find a suitable spot to erect my furnace, and to make every preparation for the arrival of my balloon from Meudon. Each day my observations contained something new either in the works which the Austrians had thrown up during the night, or in the arrangement of their forces. On the fifth day a piece of cannon had been brought to bear upon the balloon, and shots were fired at me as soon as I appeared above the ramparts. None of the shots took effect, and on the following day the piece was no longer in position. Experience enforced upon me the necessity of forming some provision against these unexpected attacks. I employed the night in fixing cords all round the middle of my balloon. Each of the aerostiers had charge of one of the ropes, and by means of them I could easily move about, and thus get myself out of range of any gun that had been trained to bear against me. I was afterwards ordered to make a reconnaissance at Mayence, and I posted myself between our lines and the enemy at half range of cannon. When the wind, which was tempestuous at first, became calmer, I was able to count the number of cannon on the ramparts, as well as the troops that marched through the streets and in the squares.

"Generally the soldiers of the enemy, all who saw the observer watching them and taking notes, came to the idea that they could do nothing without being seen. Our soldiers were of the same opinion, and consequently they regarded us with great admiration and trust. On the heavy marches they brought us prepared food and wine, which my men were hardly able to get for themselves, so closely did they require to attend to the ropes. We were encamped upon the banks of the Rhine at Manheim when our general sent me to the opposite bank to parley. As soon as the Austrian officers were made aware that I commanded the balloon, I was overwhelmed with questions and compliments.

"What causes an impression which, till one is accustomed to it, is very alarming, is the noise which the balloon makes when it is struck by successive gales of wind. When the wind has passed, the balloon, which has been pressed into a concave form by the wind, suddenly resumes its globular form with a loud noise heard at a great distance. The silk of the balloon would often burst in a case of this kind, were it not for the restraining power of the network."

After the days of Coutelle we do not read that balloons were made much use of in warfare. The only ascent in the Egypt campaign was that of a tricolor balloon thrown up to commemorate a fete. That Napoleon knew full well the value of the scientific discoveries of his time is clear from the following conversation with a learned Mohammedan, which took place in the great pyramid of Cheops:—

Mussamed. "Noble successor of Alexander, honour to shine invincible arms, and to the unexpected lightning with which your warriors are furnished."

Bonaparte. "Do you believe that that lightning is the work of the children of men? Allah has placed it in our hands by means of the genius of war."

Mussamed. "We recognised by your arms that it is Allah that has sent you—the Delta and all the neighbouring countries are full of thy miracles. But would you be a conqueror if Allah did not permit you?"

Bonaparte. "A celestial body will point by my orders to the dwelling of the clouds, and lightning will descend towards the earth, along a rod of metal from which I can call it forth."

Napoleon did not favour the use of balloons in war. Perhaps it was because he himself had such a splendid genius for war that he depended alone upon himself, and scorned assistance. Perhaps it was because if balloons were discovered to be of real utility, his enemies might make use of them as well as himself, and France retain no special advantage in them. But however this may be, on his return from Egypt he sold the balloon of Fleurus to Robertson. The company of ballooneers was dissolved, and the balloons themselves disappeared in smoke.

During the war in America, the role which the balloon played was a more important one. The Government of the United States conferred the title of aeronautic engineer upon Mr. Allan, of Rhode Island, who originated the idea of communicating by a telegraphic wire from the balloon to the camp. The first telegraphic message which was transmitted from the aerial regions is that of Professor Love, at Washington, to the President of the United States. The following is this despatch:—

"WASHINGTON, Balloon the 'Enterprise.'

"SIR,—The point of observation commands an extent of nearly fifty miles in diameter. The city, with its girdle of encampments, presents a superb scene. I have great pleasure in sending you this despatch—the first that has been telegraphed from an aerial station—and to know that I should be so much encouraged, from having given the first proof that the aeronautic science can render great assistance in these countries."

In the month of September, 1861, one of the most hardy aeronauts (La Mountain) furnished important information to General M'Clellan. The balloon of La Mountain, which arose from the northern camp upon the Potomac, passed above Washington. La Mountain then cut the cord that connected his balloon with the earth, and rising rapidly to the height of a mile and a half, he found himself directly above his enemies' lines. There he was able to observe perfectly their position and their movements. He then threw over ballast, and ascended to the height of three miles. At this height he encountered a current which carried him in the direction of Maryland, where he descended in safety. General M'Clellan was so much satisfied with the observations taken in the balloon, that, at his request, the order was given to the War Department to construct four new balloons.

If this volume of "The Library of Wonders" had not had for its single object "balloons and their history," we would have devoted a chapter to the numerous attempts made to steer balloons. We shall only say here that aerial navigation should be divided into two kinds with balloons, and without balloons. In the first case, it is limited to the study of aerial currents, and to the art of rising to those currents which suit the direction of the voyage undertaken. The balloon is not the master of the atmosphere; on the contrary, it is its powerless slave. In the second case, the discovery of Montgolfier is useless; and the question is, to find out a new machine capable of flying in the air, and at the same time heavier than the air. Birds are, without doubt, the best models to study. But with what force shall we replace LIFE? The air-boat of M. Pline seems to us one of the best ideas; but the working of it presents many difficulties. Let us find a motive power at once light and powerful (aluminium and electricity, for example), and we will have definitively conquered the empire of the air.

Advertisements in the back of the book:

Click on any items in the list below

"Charles Scribner & Co., 654 Broadway, New York, have just commenced the publication of The Illustrated Library of Wonders. This Library is based upon a similar series of works now in course of issue in France, the popularity of which may be inferred from the fact the OVER ONE MILLION COPIES have been sold."

Advertisements for books about:

acousticsanimalsarchitecturebodyegyptescapeglassheathuntingItalianMoonopticsPompeiiseastrengthsublimesunthunderwater

www.ingramcontent.com/pod-product-compliance
Lightning Source LLC
Chambersburg PA
CBHW031630210526
45464CB00004B/1830